IMAGES
*of America*

# ASHTABULA FIREFIGHTING

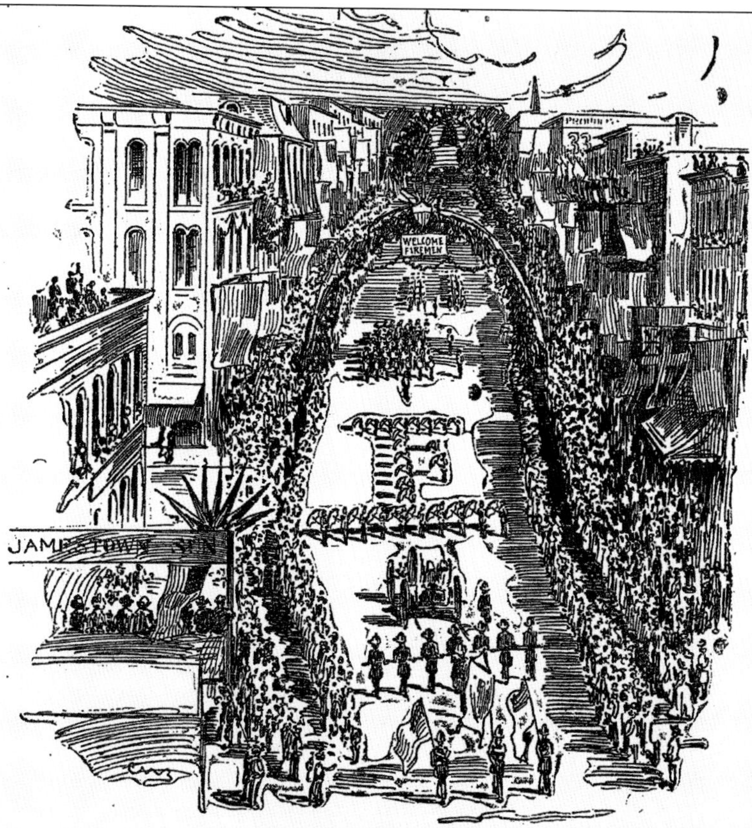

**THE PARADE.**
**Scene on Main Street. The Protections of Ashtabula, Ohio, and their Parasols.**

**THE PARADE, JULY 26, 1888.** This Jamestown (New York) *Sun* artist's sketch shows the champion "Protections of Ashtabula, Ohio, and their Parasols" as they make their way down Main Street during a firefighter's parade and drill competition in Jamestown, a small western New York city on picturesque Chautauqua Lake. Protection Fire Company minutes from August 4, 1888, feted the event as a "grand old tournament." During their 1850–1912 run as the city's leading volunteer fire company, Ashtabula's famed Protections were a prestigious, award-winning corps. Competing in local and regional drill competitions in communities including Jefferson, Painesville, Cleveland, Warren, Conneaut, Greenville, and Meadville, Protection Fire's uniformed band and drill teams amassed over 100 ribbon "premiums" for their award-winning skills. Protection's undefeated record continued until the company reached heated international competition at Buffalo, New York. Among the company's celebrated drillmasters was longtime department volunteer and 1902–1907 city fire chief Trace H. Paine. Serving the company's band and drill teams, Paine often carried two uniforms to contests. (Courtesy Ashtabula County District Library.)

*On the cover:* **PENNY FURNITURE FIRE.** Ashtabula City and Ashtabula Township firefighters struggle in vain to contain the growing blaze inside the three-story Nettleton block during the famed "Penny Furniture Company fire" of February 26, 1965. The $1 million, 10-alarm inferno, fought by 200 professional and volunteer firefighters from 10 Ashtabula and Lake County departments, destroyed downtown Ashtabula's Nettleton and Morrison Blocks and threatened the adjacent Hotel Ashtabula. (Photograph Venditti Flowers; courtesy Ashtabula Fire Department [AFD] Archive.)

IMAGES
*of America*

# ASHTABULA FIREFIGHTING

Eric A. Johnson

ARCADIA
PUBLISHING

ISBN 0-7385-4047-1

Published by Arcadia Publishing
Charleston SC, Chicago IL, Portsmouth NH, San Francisco CA

Printed in the United States of America

Library of Congress Catalog Card Number: 2006924960

For all general information contact Arcadia Publishing at:
Telephone 843-853-2070
Fax 843-853-0044
E-mail sales@arcadiapublishing.com
For customer service and orders:
Toll-Free 1-888-313-2665

Visit us on the Internet at http://www.arcadiapublishing.com

*Dedicated in loving memory of my Uncle Ronnie—veteran Cedarburg, Wisconsin, volunteer firefighter/EMT Ronald J. "Roundy" Johnson (1937–1991)—a young boy's forever hero.*

# Contents

# ACKNOWLEDGMENTS

The creative effort behind *Ashtabula Firefighting*, like most worthwhile endeavors, was accomplished with the assistance and support of many.

I am extremely grateful to the Ashtabula Fire Department (AFD) for generously opening its extensive historical archives for this book. I am particularly grateful for the support of retiring fire chief Richard F. Balog Jr. and incoming fire chief Ronald Pristera, both ardent supporters of ongoing efforts to chronicle and preserve the AFD's rich 170-year history. Their enthusiasm transformed *Ashtabula Firefighting* from a fun idea into an exciting reality.

A special thanks to "Chief Rick" for his patience, good humor, and expert technical assistance with the cd scans for *Ashtabula Firefighting.* His support of the project proved to be an invaluable inspiration.

I am also deeply indebted to engineer John Paul, the AFD's resident history buff and de facto archivist, for his invaluable assistance with accessing and organizing many of the images included herein.

My thanks also extend to the Ashtabula County District Library, the Ashtabula Great Lakes Marine and Coast Guard Memorial Museum, IAFF Local 165, the Harbor-Topky Memorial Library, the Cleveland *Plain Dealer*, and the Ashtabula *Star Beacon* for generously opening their archival collections. A special word of thanks to *Star Beacon* editor Neil Freider for generously providing access to and use of the paper's extensive archives.

I am extremely grateful to George E. Ducro III, J. Peter Ducro, Carl E. Feather, Warren Dillaway, Gail and Ron Deemer, St. Peter's Episcopal Church, Ray and Doree Petros, and Charles P. "Chuck" Brockway for their photographic contributions to *Ashtabula Firefighting.*

A heartfelt word of thanks also goes to AFD administrative assistant Kathy Gray and the numerous active and retired Ashtabula firefighters who have provided many valuable suggestions, information, and remembrances that are included throughout *Ashtabula Firefighting.*

But most of all, I offer my heartfelt thanks to my family––wife Barbara, daughter Evelyn, and son Andrew––for their loving support during my long preoccupation with researching and writing *Ashtabula Firefighting.*

# INTRODUCTION

The rich legacy of firefighting is an old tradition, one that dates back some 400 years to 17th-century Europe, where bucket brigade firefighting was considered a primary civic obligation of every able-bodied male. The firefighting tradition would cross the Atlantic Ocean to the United States with the formation of the first fire company in Boston in 1678. During the 17th, 18th, and 19th centuries, informal citizen bucket brigades would eventually give way to the creation of thousands of better-equipped volunteer civic fire companies and, in larger communities, paid professional fire departments.

In the pioneering communities of northeast Ohio's "Western Reserve," community firefighting was an informal enterprise. Fires, a frequent occurrence, were fought by untrained civilians manning bucket brigades. The tedious, time-consuming, and labor-intensive enterprise was not intended to save the burning structure, but rather served as a defense against a larger conflagration that might spread to adjacent structures and perhaps ignite whole neighborhoods or the entire community.

Founded in 1803 and incorporated as a village in 1831, Ashtabula steadily grew in size and stature. And as the village grew, so did its need for an organized volunteer fire department to serve local residents.

The establishment of a volunteer fire corps became a reality on November 29, 1836, when Ashtabula's village council and Mayor H. T. Rees established the pioneering Eagle Fire Company and approved the $700 purchase of a hand-operated William Platt and Company pumper engine. By early 1837, contractor S. S. Fassett had built a $115 engine house on the northwest corner of Main Avenue and Fisk (West 48th) Street on land donated by I. H. Fisk. The village appointed Isaac Lockwood to oversee the pumper, with Amos C. Hubbard serving as Eagle's company foreman.

But the ambitious venture proved short-lived, almost ending before it began. The crude hand-pumper, one of the first ever made in the United States, quickly broke down and was soon declared a hopeless case. Reported the Ashtabula *Beacon-Record*, "In the course of a few months it became out of repair. It was tinkered by several until it was ultimately tinkered to death and became utterly useless."

Ashtabula residents would later band together again to provide organized volunteer firefighting with the creation of the short-lived Washington Company.

Home to 2,177 residents by mid-century, Ashtabula's volunteer firefighting efforts took on a solid foundation with Fred W. Blakeslee's July 6, 1850, establishment of the city's famed "Ever Ready" Protection Fire Company during the administration of village mayor Lorenzo Tyler.

A new hand-pumper engine, which tapped massive 2,000-barrel fire cisterns dug around the village, was ordered from a Rochester, New York, fire apparatus manufacturer. With its massive five-foot-diameter wheels, Protection's pumper was transported by firefighters pulling long ropes.

Recalled longtime Protection volunteer A. L. Gregory in a 1950 *Star Beacon* feature, "It took some pretty good sprinting to outrun Horace Streeter and Jim Reed. Jim Reed always led the fire runs."

Charter Protection Fire officers were foreman N. W. Thayer, first assistant foreman George Russell, second assistant Mason King, treasurer J. H. Mather, and secretary Alford Hendry.

In addition to Thayer, other Protection Fire foremen over the company's 1850–1912 firefighting run included W. H. Morrison, George W. Lockwood, John C. Crosby, George Brake, village mayors William Clover and William Ward Smith, Seth Warren, Charles W. Savage, A. D. Shilinger, James Reed, W. H. Hubbard, W. B. Hubbard, E. G. Dickinson, William Sievers, A. T. Faulkner, John P. Ducro II, and future Ashtabula fire chiefs Fred W. Blakeslee, G. A. Knapp, Trace H. Paine, and George E. Ducro Sr.

Other Protection Fire alumni to later serve as Ashtabula's fire chief included Henry H. Hall and Ray N. Mahan (1929–1939).

Unlike Eagle Fire, Protection Fire enjoyed good luck and a solid footing. Observed the *Beacon-Record*, "This company kept its organization and equipment in good condition for effective service . . . The company embraced among its members the leading men of the town in active life, and did very effective work and in several instances saved the town from a big conflagration."

Protection Fire founder Blakeslee, later appointed fire chief by mayor Lorenzo Tyler, served a distinguished eight-year tenure as chief, a period marked by his development of a well-organized volunteer department and the establishment of three additional volunteer fire companies—*c.* 1868 Alert Hook and Ladder, Lake Erie Hose Company, and the *c.* 1874 Neptune Hose Company.

By the early 1900s, as the nation's golden age of volunteer firefighting drew to a close, the Ashtabula Fire Department operated eight neighborhood-based volunteer fire companies across the city, each staffed with 18 men and three executive officers. Those companies were as follows: Protection Fire, Main Avenue; (John S.) Bonnar Hose, Lake and Depot Streets; Neptune Hose, Center Street; Topky Fire, Bridge Street; F. C. Moore Hook and Ladder, Main Avenue; Raser Hose, Bridge Street; Nickel Plate Hose, Lockwood (West 50th) Street; and (Russell C.) Humphrey Hose, serving Ashtabula's old "Swede Town" section from its station on Pacific (East 15th) Street.

Recalled Protection Fire volunteer A. L. Gregory of horse-drawn fire runs, "When the big fire bell in City Hall sounded the alarm, the volunteers dropped what they were doing and ran . . . By the time the firemen had assembled, the drop harnesses—always hung in readiness—were in place on the horses. And then the fire engine, with bell clanging and black smoke sprouting from the stack, would speed to the fire, trailed by small boys and watched by gaping bystanders."

In the late 1800s and early 1900s, Ashtabula prospered into a booming Lake Erie port and manufacturing center, incorporating as a city in 1891. With Ashtabula's rising stature and economic expansion, the city's population grew significantly, rising from 3,394 residents in 1870 to 12,949 by 1900, growth aided in part by the annexation of two rival communities—Ashtabula Harbor in 1877 and Swede Town in 1890. With its industrial expansion and eye-popping 30 percent population growth between 1900 and 1910, the need for a professional paid firefighting force became evident as Ashtabula's population quickly rose to 18,266.

During the waning months of the 1908–1911 administration of Republican mayor H. D. Cook, the Ashtabula city council, at the urging of safety director Dr. I. M. Hadcock and fire chief George E. Ducro Sr., transformed Ashtabula's long-standing volunteer fire department into a paid, full-time professional corps. With the October 25, 1911, transformation of the Ashtabula Fire Department into one of Ohio's first paid fire departments, the city's volunteer fire companies began to disband, ending a colorful era of Ashtabula firefighting history. Idled as a firefighting corps in 1912, Protection Fire continued as a fraternal social organization until disbanding in May 1918.

A six-man crew, largely longtime department veterans, comprised the city's initial professional firefighting roster—Capt. William L. Haigney, engineer Charles L. Devereaux, and firemen Fred D.

Alexander, Charles E. Lasky, John E. Wylie, and Daniel Condron. The crew was based in the former Topky Fire quarters inside the Harbor Fire Station at the old Harbor City Hall on Bridge Street. Ashtabula's professional fire department expanded operations in February 1912 with a six-man paid professional fire company at the city's newly built $7,499 headquarters Uptown Fire Station on Park Avenue at West 48th Street in downtown Ashtabula. Six of the city's eight volunteer companies disbanded, with Nickel Plate Hose and Humphrey Hose maintained into the 1920s to augment professional firefighting services in far-flung west- and east-side areas of the city.

In the early days of the paid professional fire department, each station, by city ordinance, was staffed by six paid men––a captain, an engineer, and four firemen. Both crews were backed by emergency call lists of 10 volunteers remunerated for time on duty. Council-approved annual salaries for the inaugural paid fire department included $400 for fire chief Ducro, $1,000 for captains and engineers, and $900 for firemen.

Prior to the implementation of a two-platoon system in 1919, the department's two stations were each staffed by six-man crews that worked a staggered schedule of five 24-hour days (with three home meal breaks), followed by one day off––a staggering 120-hour work week something akin to indentured servitude.

With the February 1912 department expansion, roster assignments included George H. Henniger, William L. Haigney, Charles E. Lasky, John Wylie, Daniel Condron, and Luke J. Kelly at the Harbor Fire Station, with Charles L. Devereaux, Archie Castle, Fred D. Alexander, John L. Huggins, Scott Leahy, and James Dunlavy posted to the headquarters Uptown Fire Station in downtown Ashtabula.

Concurrent with the historic 1911 organization of Ashtabula's paid fire department, Hadcock and Ducro also oversaw the trailblazing purchase of the department's first two pieces of motorized "auto fire truck" apparatus, with the *Beacon-Record* noting that such equipment had successfully "passed beyond the experimental stage and become a reality" in larger regional cities like Detroit. Noted Ducro of the improvements, "Anyplace horses can go, the Auto will go––and much faster."

But city missteps and council squabbling nearly killed the fledgling paid professional fire department. Ashtabula's "nearly penniless" firefighters threatened to quit en masse in late December 1911 after the city failed to pay them their initial two months of wages because it had failed to pass an ordinance setting salaries when it created the paid department. Complicating matters, there was also considerable contention as council "came to loggerheads" over the monthly salary rate being set at either $70 or the $75 originally promised by Mayor Cook. A city emergency services crisis was averted when a wealthy anonymous citizen stepped forward to advance salary payments to city firefighters in the interim, with firefighter pay vouchers used as security collateral for the loans.

With the creation of the paid professional fire department, all qualified men took a civil service examination for their job classifications with the exception of part-time fire chief Ducro, who had been appointed to the post in 1907.

With Ducro's 1916 resignation after the city adopted its current city charter, the Ashtabula Fire Department operated 13 years without a fire chief, headed instead by two fire captains, with the city manager acting as ex-officio fire chief. Following an expansion of equipment, facilities, and personnel, the need for a full-time paid fire chief became evident. In July 1929, city manager Fred R. Hogue appointed 15-year Protection Fire volunteer Ray N. Mahan as the city's first full-time fire chief.

After Mahan's 1929–1937 tenure, civil service went into effect for the position of fire chief, beginning with department veteran Charles L. Devereaux in July 1939. After Devereaux's May 1943 death following a protracted illness, longtime city firefighter Fred D. Alexander served a brief seven-month tenure as chief. With Alexander's May 1944 death from a heart ailment following a brief illness, veteran firefighter Adolph T. "Adie" Rohl became fire chief in 1945, serving until his 1963 retirement.

Subsequent fire chiefs that have risen through the department's ranks under civil service include James M. Doran (1963–1971), Charles L. Mosier (1971–1987), Norman Jepson (1987–1997), Richard F. Balog Jr. (1997–2006), and Ronald Pristera (2006– ).

With the exception of the enduring close-knit brotherhood bonds that continue to bind firefighters together, today's modern 27-person, three-shift Ashtabula Fire Department bears little resemblance to the fledgling 12-man department of 1911–1912. Exciting changes over the years include the regular addition of state-of-the-art firefighting apparatus, the proactive 1967 creation of a full-time fire prevention bureau, and the development of new city fire stations, including the enduring *c.* 1975 headquarters Station One facility on Main Avenue in downtown Ashtabula.

In this, the Ashtabula Fire Department's 170th anniversary year, it is with great pleasure that I present *Ashtabula Firefighting.*

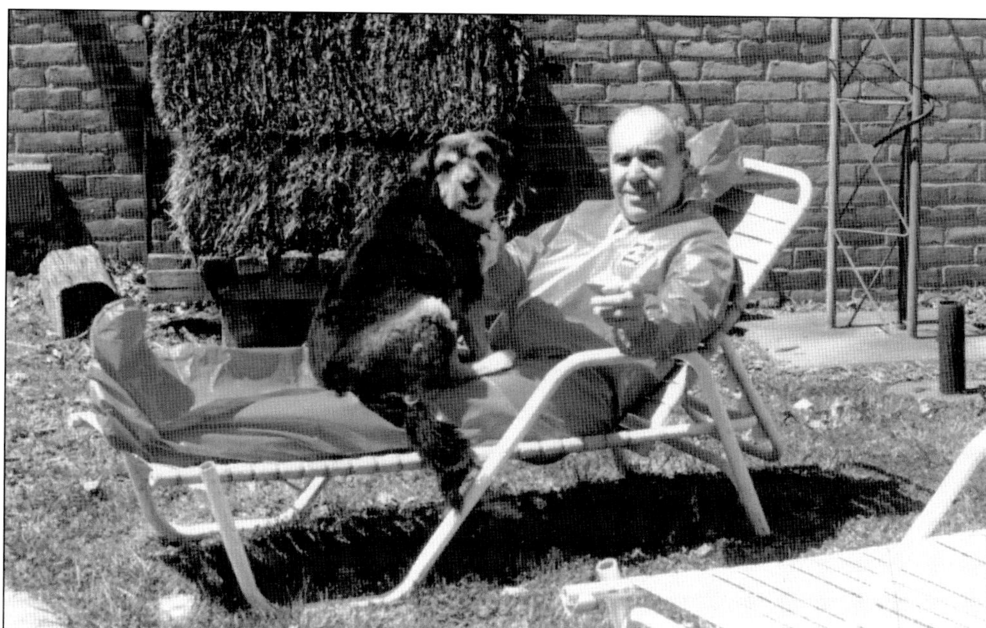

**CAPT. THOMAS J. LAVECK AND SCRUFFY, 1998.** Taking a break during a Station One haz-mat (hazardous materials) training session, 33-year department veteran Capt. Thomas Laveck (1967–2000) relaxes with the Ashtabula Fire Department's most recent canine mascot, the late Scruffy. Firefighters adopted the lost dog after it turned up on their doorstep in the mid-1990s. Of the Station One roster, Scruffy took a special shine to Laveck, a St. John's High School alumnus and part-time Ashtabula Harbor barber (1960– ). Recalls Laveck of Scruffy, "He stuck to me like glue for some reason." Scruffy was later adopted by firefighter Ron Deemer. (Courtesy Gail and Ron Deemer.)

# *One*

# The Volunteer Era

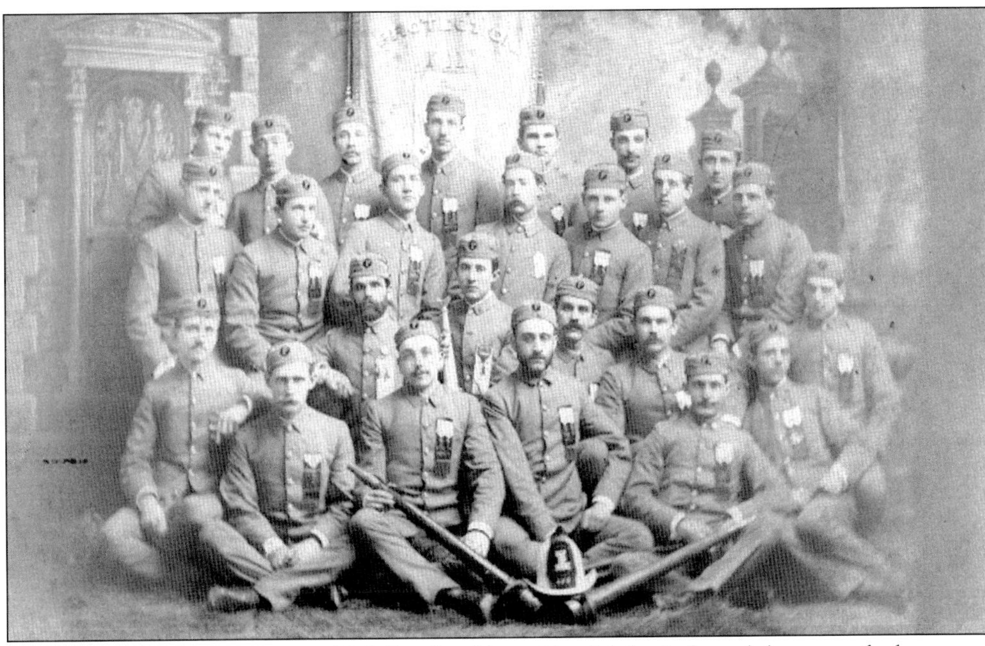

**Protection Fire Company, 1885.** Feted as "the pride of Ashtabula and the envy of other towns for miles around," the volunteer Protection Fire Company was established on July 6, 1850, by future eight-year village fire chief Fred W. Blakeslee and 35 like-minded men at an organizational meeting held at the home of village mayor Lorenzo Tyler. Operating under the "Ever Ready" motto and regarded as the aristocrat of Ashtabula's volunteer fire companies, Protection Fire's roster included many of the city's prominent businessmen and civic leaders during its 1850–1912 firefighting run. Pictured on Protection Fire's 1885 roster are, from left to right, (first row) William Sievers, William Morrison, James Reed, John Robinson, and unidentified; (second row) Fred Wood, John Stevenson, Charlie Savage, Horace Streeter, Gene Henry, and Amos Hubbard; (third row) James Bailey, Will Hubbard, future city fire chief Trace H. Paine, Jimmy Biggers, foreman John P. Ducro, John Hubbard, and George Hubbard; (fourth row) ) Ed Dickinson, John Falkner, Frank Nearing, Fred R. Morrison, Frank Silliman, Stanley Amsden, and John Topky. (Courtesy AFD Archive.)

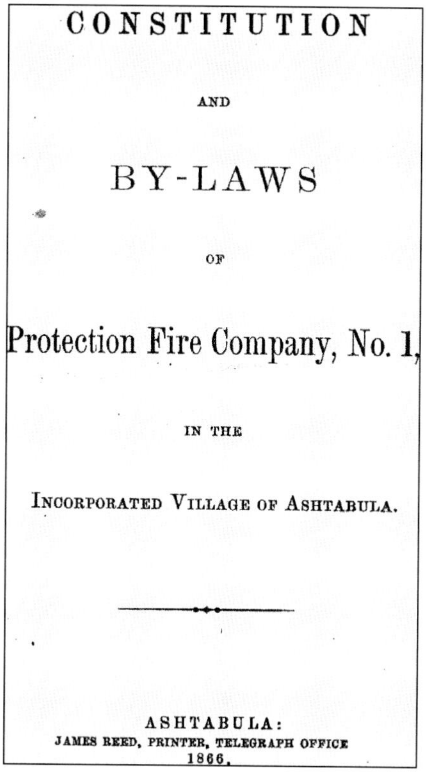

# CONSTITUTION

AND

# BY-LAWS

OF

## Protection Fire Company, No. 1,

IN THE

INCORPORATED VILLAGE OF ASHTABULA.

ASHTABULA:
JAMES REED, PRINTER, TELEGRAPH OFFICE
1866.

**PROTECTION FIRE CONSTITUTION AND BY-LAWS, MARCH 1866.** This booklet is the earliest-known Ashtabula firefighting artifact. Protection Fire officers were foreman G. A. Knapp, first assistant chief E. A. Hitchcock, second assistant chief James H. Savage, secretary A. S. Bouker, and treasurer Edward H. Fitch. Officers were elected annually by member ballots, cast on the first Wednesday in April. Monthly drills were held on first Wednesdays at 5:00 p.m. between April and October. (Courtesy J. Peter Ducro.)

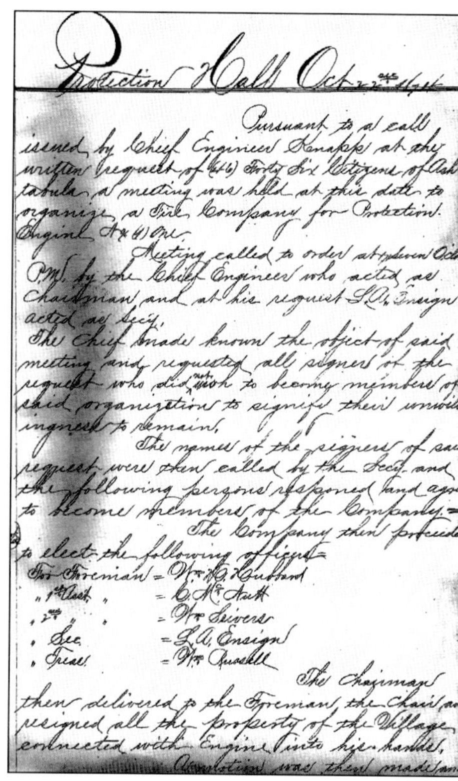

Protection Hall Oct [illegible]

Pursuant to a call issued by Chief Engineer Knapp at the written request of (46) Forty Six Citizens of Ashtabula a meeting was held at this date to organize a Fire Company for Protection Engine No (1) One.

Meeting called to order at seven o'clock P.M. by the Chief Engineer who acted as Chairman and at his request L.A. Ensign acted as Secy.

The Chief made known the object of said meeting and requested all signers of the request who did not wish to become members of said organization to signify their unwillingness to remain.

The names of the signers of said request were then called by the Secy and the following persons responded and agreed to become members of the Company.

The Company then proceeded to elect the following officers

For Foreman = Wm H. Hubbard
" 1st Asst " = C. McNutt
" 2nd " " = Wm Seivers
" Sec = L.A. Ensign
" Treas = Wm Russell

The Chairman then delivered to the Foreman, the Chair and resigned all the property of the Village connected with Engine into his hands.

A motion was then made [illegible]

**PROTECTION FIRE MINUTES, APRIL 1876.** This Spencerian Script entry in Protection Fire's leather-bound 1871–1918 journal details the company's 1876 annual meeting at Protection Fire Hall. New volunteer firefighters were Albert Gaston, J. S. Burrows, Henry Aberle, W. H. Hubbard, and E. E. Russell. Elected as 1876–1877 company officers were foreman W. H. Hubbard, first assistant foreman George Brake, second assistant foreman L. A. Ensign, secretary D. H. Collins, and William Russell, treasurer. (Courtesy AFD Archive.)

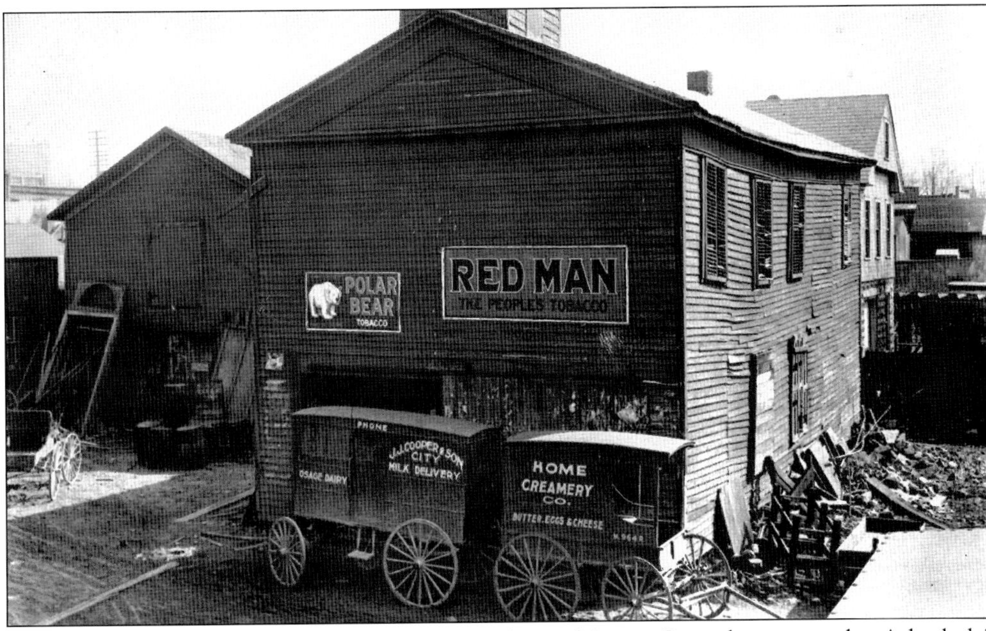


**ASHTABULA VILLAGE HALL.** Prior to 1882, this relocated Center Street livery served as Ashtabula's village hall. Originally located on West Forty-fourth (Division) Street near Main Avenue, Ashtabula's village hall also housed Protection Fire Company. Fire hose was hung to dry in the rooftop cupola, while the ground floor housed Protection's sizeable *c.* 1850 hand-pumper, which tapped 2,000-barrel fire cisterns dug at various locales around the village. (Courtesy Ray and Doree Petros.)

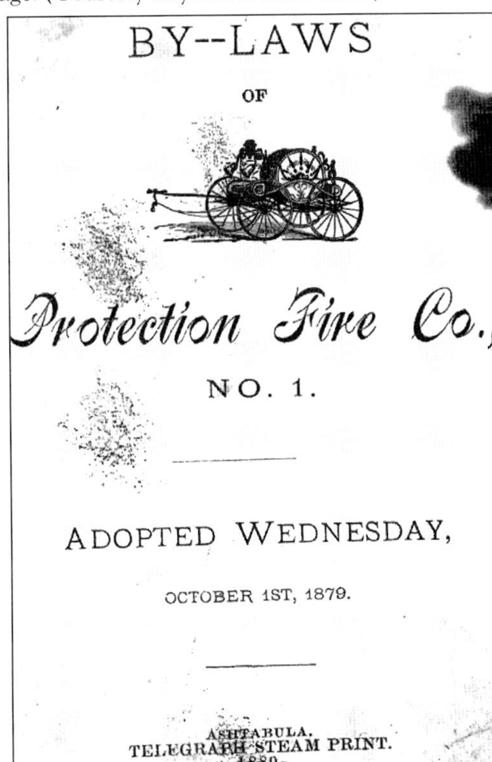
BY--LAWS

OF

*Protection Fire Co.,*

NO. 1.

ADOPTED WEDNESDAY,

OCTOBER 1ST, 1879.

ASHTABULA.
TELEGRAPH STEAM PRINT.
1880.

**PROTECTION FIRE COMPANY BY-LAWS, OCTOBER 1879.** Village mayor Henry L. Morrison and a by-laws committee of John W. Collins, John W. Stevenson, and W. B. Menneiley updated the village's fire department ordinances in 1879. Protection Fire, Lake Erie Hose, Neptune Hose, and Alert Hook and Ladder volunteers were remunerated for firefighting and training at the rate of 25¢ an hour, paid from the village's fire and water fund. (Courtesy AFD Archive.)

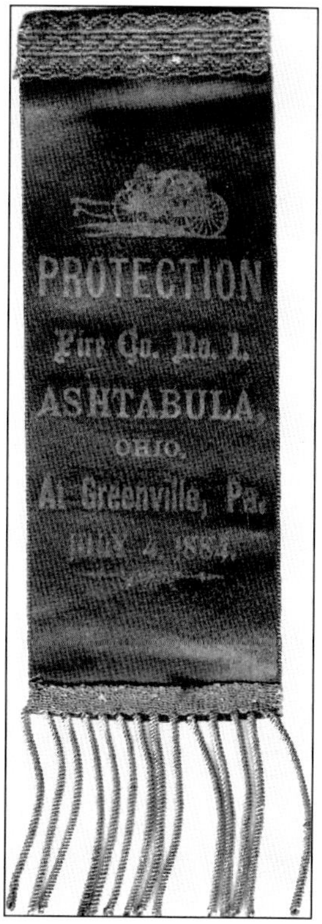


**COMPETITION "PREMIUM," JULY 4, 1884.** Feted as a "crack drill team," Ashtabula's award-winning Protection Fire Company set the standard in the firefighting parade, band, and drill competitions popular in the 1880s. Ranking as one of the most-decorated fire companies in Ohio, the region, and the nation, Protection Fire won over 100 ribbon premiums, including this red-and-gold award ribbon from an 1884 Independence Day competition in Greenville, Pennsylvania. (Courtesy AFD Archive.)

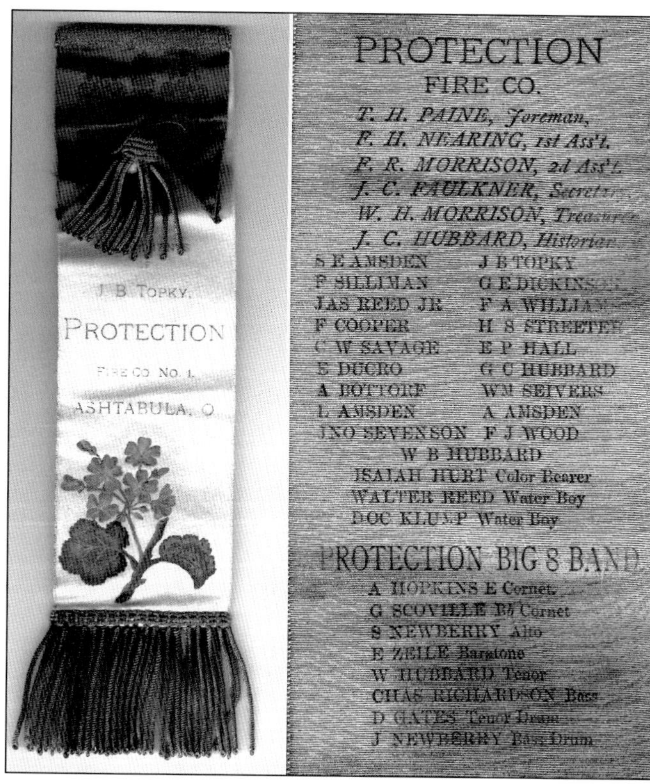


**PROTECTION FIRE COMPANY COMPETITION RIBBON, JULY 4, 1887.** Presented compliments of John B. Topky, a prominent volunteer firefighter and Harbor businessman, this ribbon and others like it were awarded by Ashtabula's "Ever Ready" Protection Fire Company as it hosted a drill, parade, and band competition for city fire companies and visiting departments from Painesville, Jefferson, Warren, Meadville, and Greenville. The reverse side (right) lists Protection's 1887 roster. (Courtesy AFD Archive.)

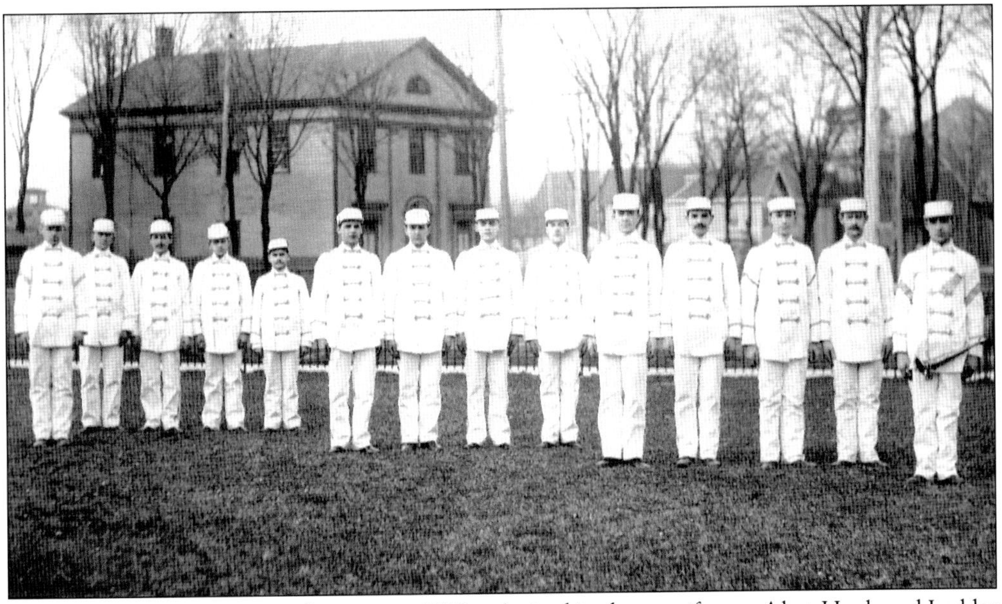

**ALERT HOOK AND LADDER COMPANY, 1880S.** Attired in dress uniforms, Alert Hook and Ladder volunteers gather at downtown's North Park. Among those pictured are Arthur J. Richardson; Fred Dickenson; A. J. Dittenhaver; Ollie Topky; Ed Redhead; George Gaskell; Clarence Case; Silas Chapel; Ned Pierce; John P. Hubbard; and Milton Soules. Alert, headquartered at Depot (West Thirty-second) Street and Lake Avenue, was organized in 1868. (Courtesy Ray and Doree Petros.)

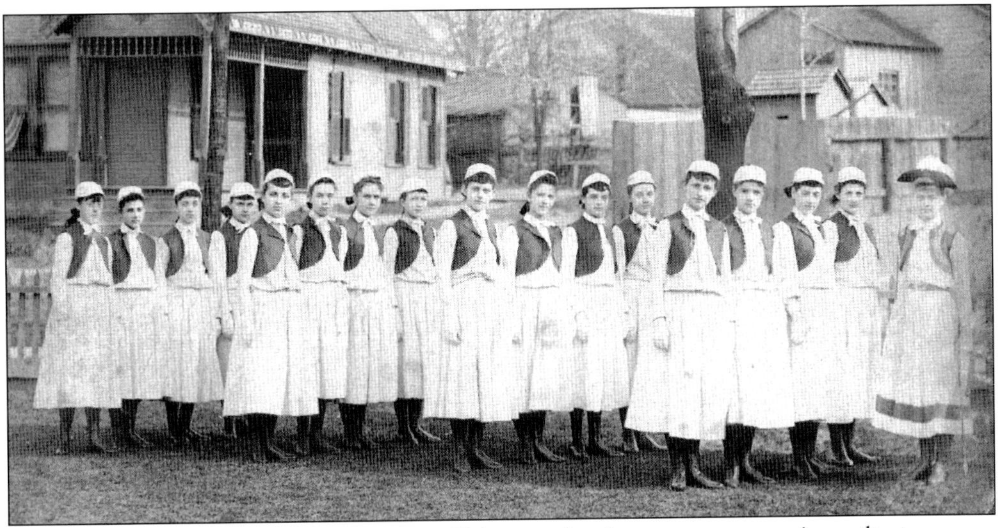

**YOUNG LADIES' DORCAS SOCIETY, 1888.** Volunteer fire department women's auxiliaries are an old tradition, as evidenced by Protection Fire's red-vested Dorcas Society "drill and parade girls." Pictured from left to right are as follows: Ella Hoyt Kelly, Anna Burnett, Emily Jaques, Nora Haskell, Mary Case, Mary Stevenson-Dickenson, Bertha Clark, Magnee Case, Laura Chisholm, Julia Cooper Topky, Catharine "Kate" Carlisle, Lucy Tombes, Dot Stewart, Bertha Webb, Ella Mygatt, unidentified, and "General" Bessie R. Carlisle. (Author's collection.)

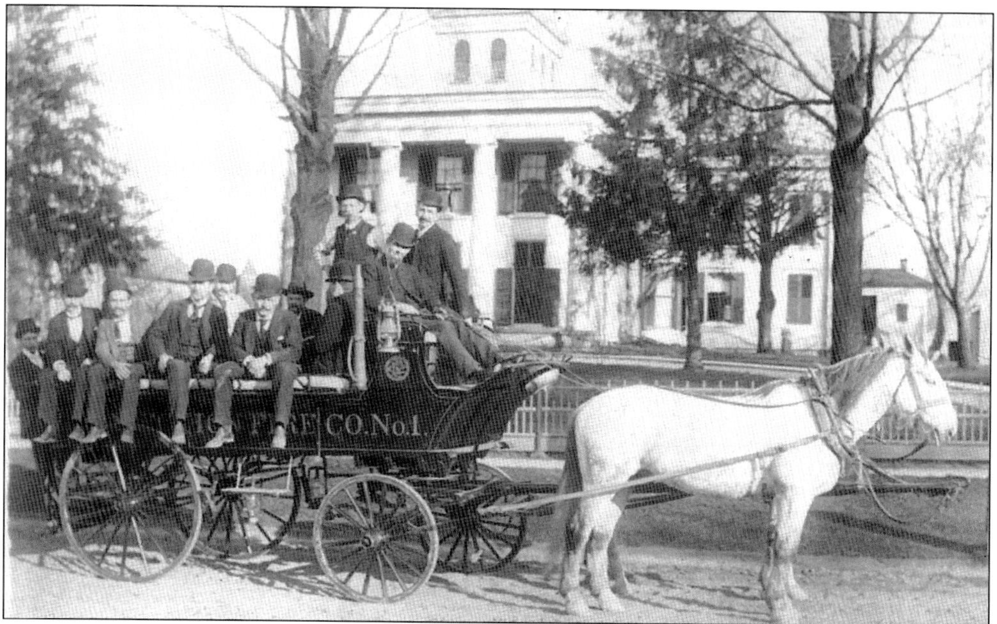


**PROTECTION FIRE COMPANY, 1889.** Protection Fire volunteers are on their hose wagon outside the William H. Morrison home, 500 West Prospect Avenue. Pictured from left to right are Tom Hall, Ed Hall, John Newberry, Trace Paine, William Morrison, John P. Ducro, Lewis Amsden, A. L. Gregory, William Sievers, Frank Gregory, and F. R. Morrison. Duties of Protection, Lake Erie Hose, and Alert Hook and Ladder volunteers included staging monthly summertime parades and band concerts. (Courtesy AFD Archive.)

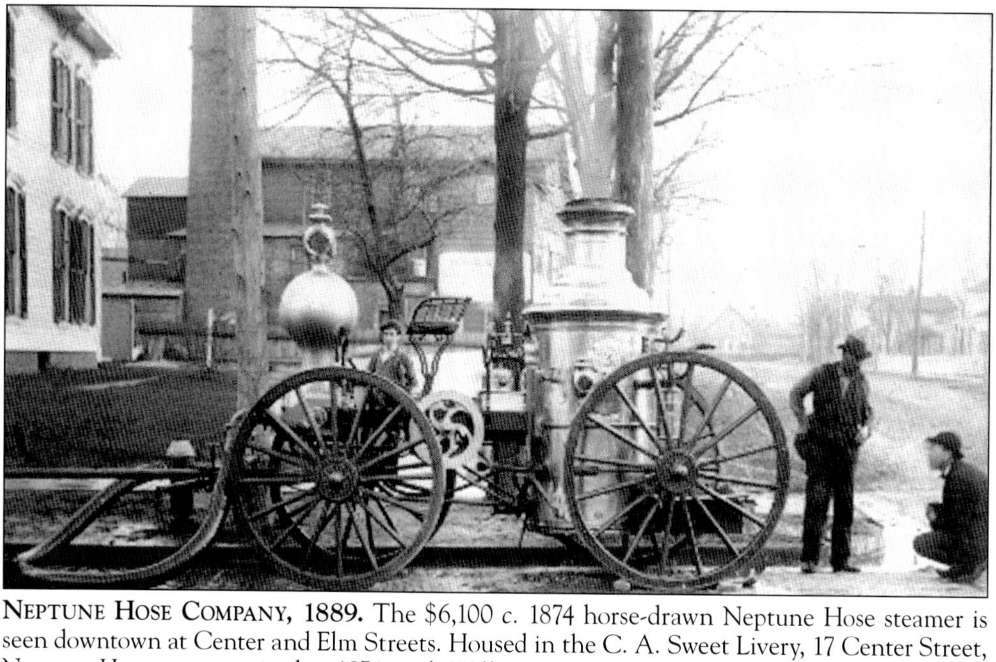

**NEPTUNE HOSE COMPANY, 1889.** The $6,100 c. 1874 horse-drawn Neptune Hose steamer is seen downtown at Center and Elm Streets. Housed in the C. A. Sweet Livery, 17 Center Street, Neptune Hose was organized in 1874, with William Sanderson appointed as company engineer at an annual salary of $150. By 1878, 34 volunteers under Capt. Paul Ford and foreman B. H. Rickford manned Neptune's steamer and 1,100 feet of hose. (Courtesy AFD Archive.)

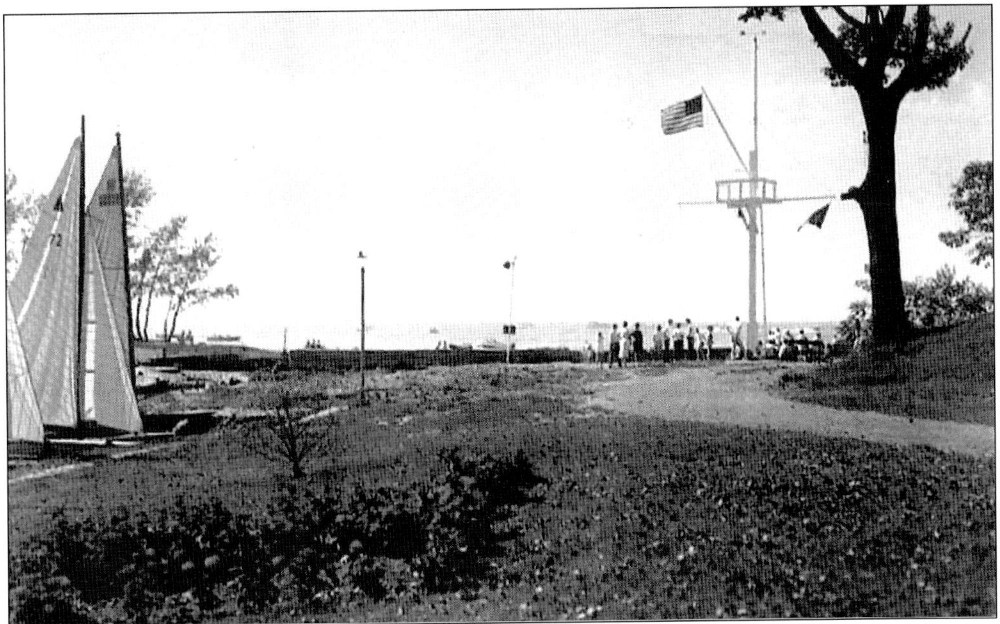

**REDBROOK.** Part fire corps, part fraternal organization, Protection Fire in 1890 allocated $150 to purchase a private c. 1884 lakefront resort in Saybrook Township along Redbrook Creek, today's 250-member Redbrook Boat Club, 4645 Lake Road West (pictured in the 1950s). Hosting company socials and picnics on the lakefront grounds, Protection's since-razed Redbrook Cottage, painted in company red and white, was open to active and retired firefighters for three-day summertime vacations. (Author's collection.)

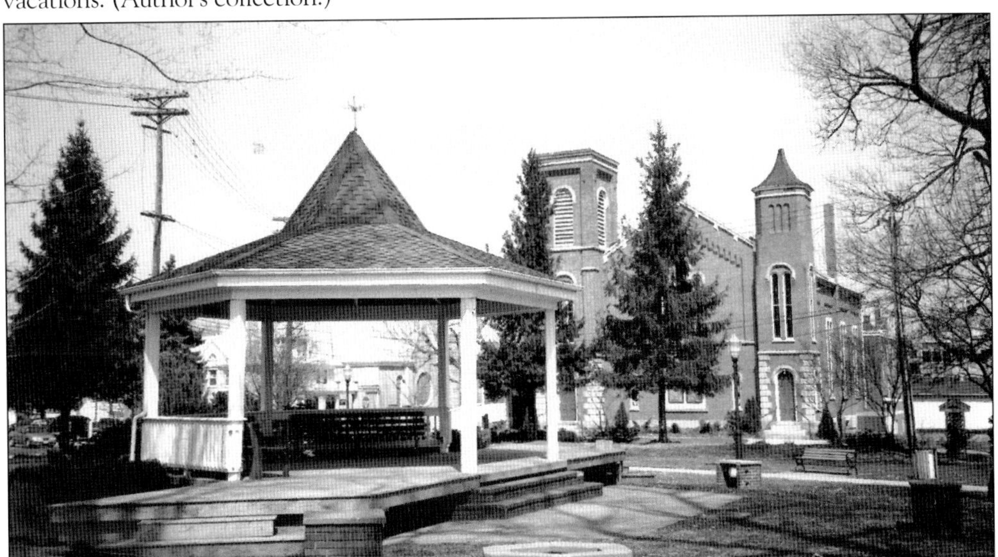

**NORTH PARK, 2006.** Originally Hall Smith Park, downtown's scenic North Park has long been a center of civic life. Pictured is the popular c. 1988 bandstand, built in memory of Ashtabula's retailing Carlisle family. North Park's original c. 1868 bandstand hosted monthly summertime band concerts by the city's Protection Fire, Lake Erie Hose, and Alert Hook and Ladder volunteers, staged in conjunction with their first-Wednesday-of-the-month drills and parades. (Author's collection.)

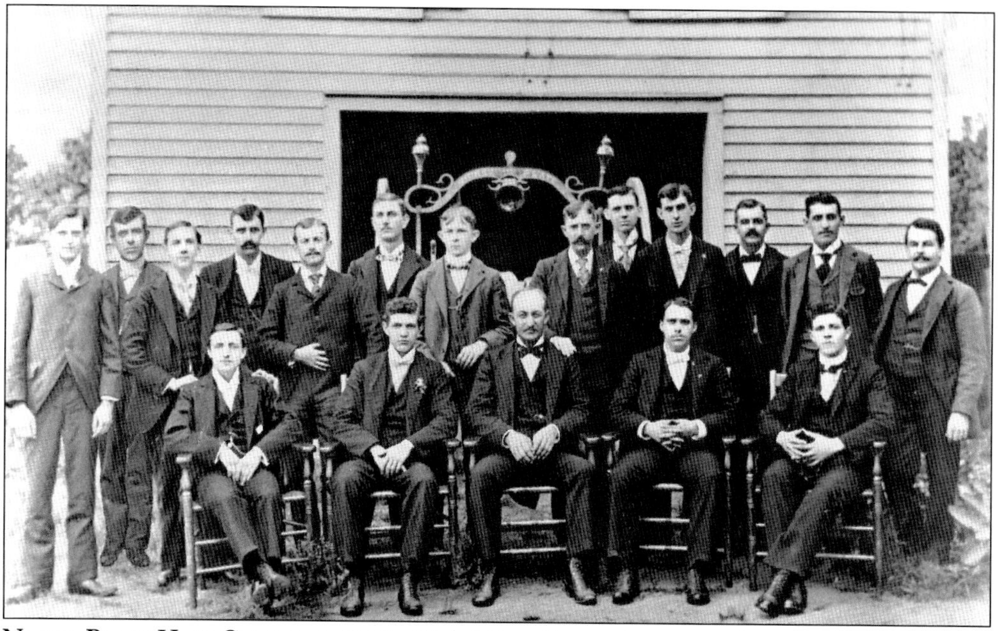

**NICKEL PLATE HOSE COMPANY, 1880S.** Nickel Plate Hose volunteers gather outside their quarters at 39 Lockwood (West Fiftieth) Street, between Cornell and West Avenues. Named for the nearby "Nickel Plate" Railroad, Nickel Plate Hose served Ward 5. Following the 1911–1912 creation of the paid professional fire department, volunteer Nickel Plate Hose was maintained into the early 1920s to provide enhanced fire service to the city's far west side. (Courtesy AFD Archive.)

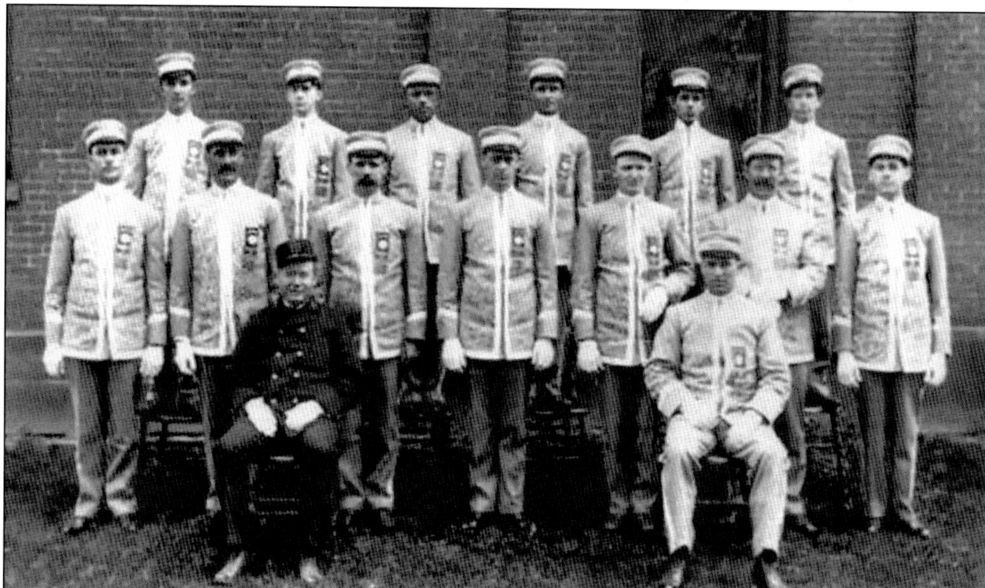

**F. C. MOORE HOOK AND LADDER COMPANY, 1890S.** The smartly attired volunteer firefighters comprising F. C. Moore Hook and Ladder pose for posterity in their dress uniforms, reserved for public functions including the firefighter parades, competitions, and band concerts popular at the time. Established in 1890 as successor to the disbanded Alert Hook and Ladder, Moore officers by 1902 included foreman Arthur J. Richardson and secretary/treasurer John P. Hubbard. (Courtesy AFD Archive.)

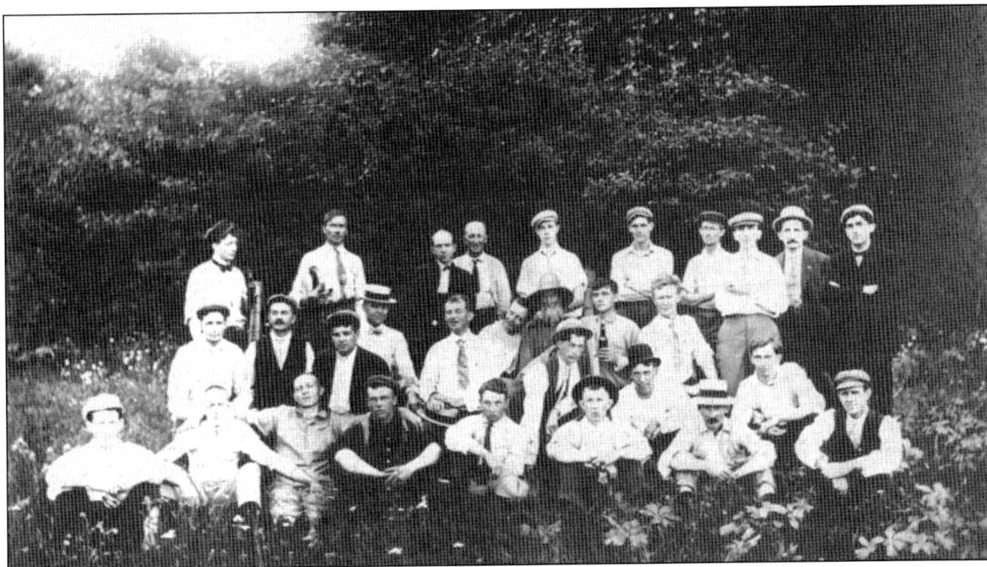

**West End Fire Company.** Volunteers comprise Ashtabula's short-lived West End Fire picnic at Fobes Flat. Shown here from left to right are (first row) unidentified, H. Cambet, Miles Moody, John Johnson, E. H. Tyler, unidentified, Dalis Whitaker, ? Berridge, unidentified, ? Berridge, and John Strubbe; (second row) ? McCarroll, George and Charles Flint, Fred Topper, Plin Lovejoy, three unidentified, and Lou Strong; (third row) unidentified, D. Weeks, F. Redman, S. Lovejoy, ? Luce, Ralph Luce, Sherm Luce, Charles Mitchell, Frank Whitacre, and Glen Wood. (Courtesy AFD Archive.)

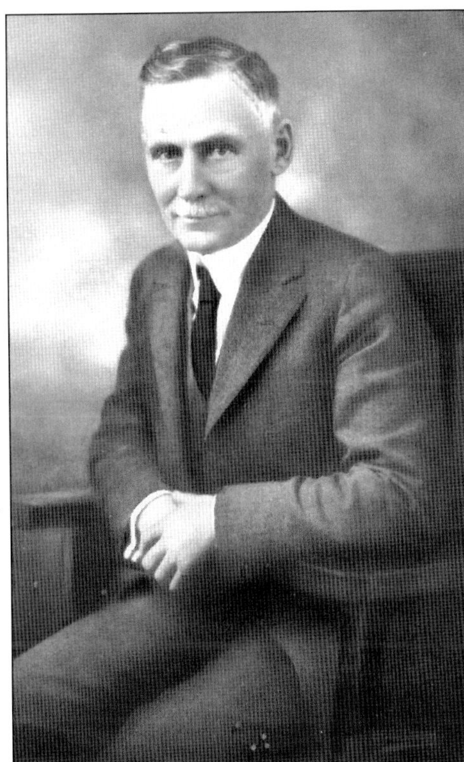

**John P. Ducro.** Businessman John P. Ducro II (1865–1950) was a prominent longtime Protection Fire volunteer firefighter and company officer, serving as foreman. Long associated with the family's enduring c. 1853 undertaking business, Ducro also served as director of the People's Savings and Loan and was the founder of the Commercial Bank. Active in other civic affairs, Ducro established the Ashtabula Auto Club to lobby for local road improvements. (Courtesy J. Peter Ducro.)

CHAPTER IX.

*FIRE DEPARTMENT AND FIRES.*

SECTION
121 Fire Department, how composed, officers, their salaries and compensation.
122 Chief and Assistants appointed by Mayor,
123 Powers of Chief and his duties defined.
124 Duties of First and Second Assistants.
125 Duty of Foreman.
126 Duties of First and Second Assistant Foreman.
127 Duty of Secretary.

SECTION
128 Duty of Treasurer.
129 Election of other officers.
130 Companies may make by-laws with approval of Council.
131 Suspension and discharge of members.
132 Officers shall not incur liability against City.
133 Council shall make appropriations for rent, light and fuel.
134 Companies shall not incur liability for rent and fuel.

How constituted.

Election by companies.

SECTION 121. The Fire Department shall consist of one Chief of Department, one First Assistant Chief and one Second Assistant Chief, for the city at large; eight volunteer companies of eighteen men each as follows: Protection Hose Company, Neptune Hose Company, Nickle Plate Hose Company, Bonner Hose Company, Topky Hose Company, Raser Hose Company, Humphrey Hose Company and Moore Hook and Ladder Company; each of which said companies shall at its first regular meeting in April of each year, elect from its members the following officers, to-wit: One Foreman, one First Assistant Foreman, one Second Assistant Foreman, one Secretary and one Treasurer; and each of said companies may by its by-laws or rules create such other officers as it may desire; the officers of the city at large shall be paid annual salaries as

— 50 —

**REVISED ORDINANCES OF THE CITY OF ASHTABULA, MARCH 1900.** By city ordinance, the Ashtabula Fire Department encompassed eight companies of 18 volunteers. Annual department salaries included $150 for the fire chief. Volunteer firefighters were paid for service rendered, 75¢ for the first hour, 50¢ an hour thereafter. Training drills were paid at 25¢ an hour. Each fire company received $140 annually for rent and utilities. (Courtesy AFD Archive.)

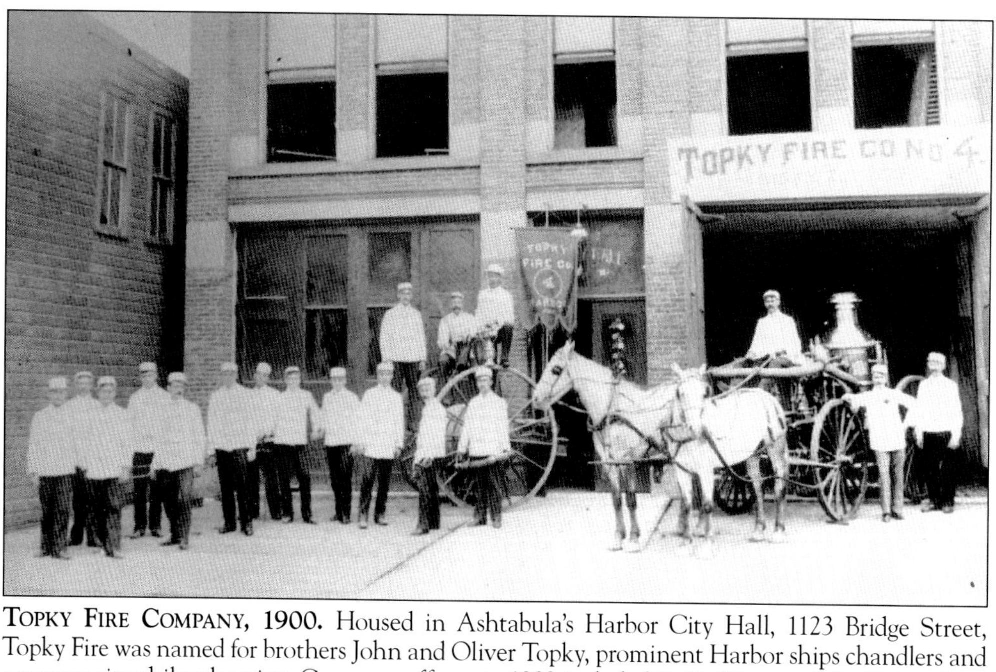


**TOPKY FIRE COMPANY, 1900.** Housed in Ashtabula's Harbor City Hall, 1123 Bridge Street, Topky Fire was named for brothers John and Oliver Topky, prominent Harbor ships chandlers and generous city philanthropists. Company officers in 1902 included C. H. Watrous, John F. Condron, Reuben E. Bugbee, W. C. McCreary, and P. H. O'Brien. A 1906 expansion of Harbor City Hall created today's familiar red brick and terra-cotta facade. (Courtesy Ray and Doree Petros.)

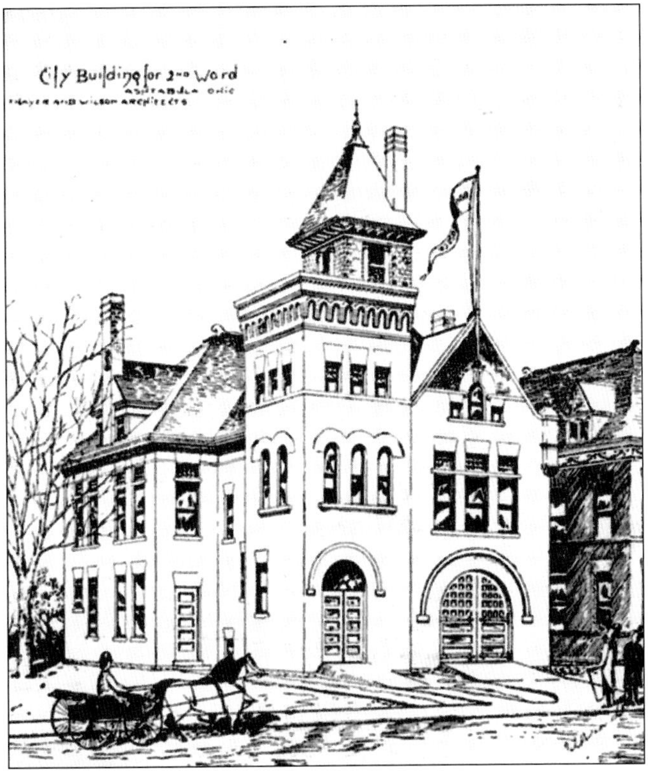


**SECOND WARD CITY HALL, LAKE AVENUE AND DEPOT STREET, 1902.** Ashtabula's Second Ward City Hall housed the volunteer Bonnar Hose Company. Designed by Ashtabula architects Thayer and Wilson, Second Ward City Hall was built in the early 1890s and razed in July 1945. The tower housed a c. 1854 fire bell cast by Meneeley's (based in West Troy, New York), used by Lake Erie Hose firefighters and their Bonnar Hose successors. (Courtesy AFD Archive.)

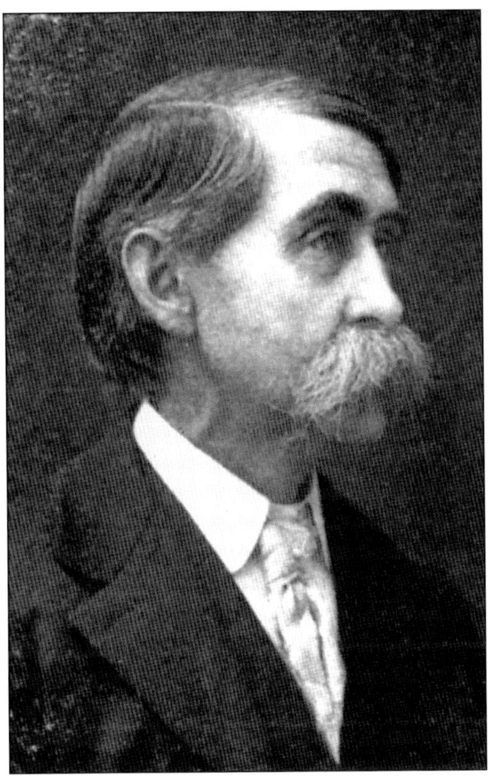

**FRED W. BLAKESLEE, 1902.** Professional photographer Fred W. Blakeslee was feted as "true to every obligation, fair and just in all of his dealings, and capable and progressive in his ideas." Organizing Protection Fire Company in 1850, Blakeslee laid the foundation for today's Ashtabula Fire Department. Fire chief for eight years and a longtime Protection Fire volunteer, it was said "the town never had a better fireman than Mr. Blakeslee." (Courtesy AFD Archive.)

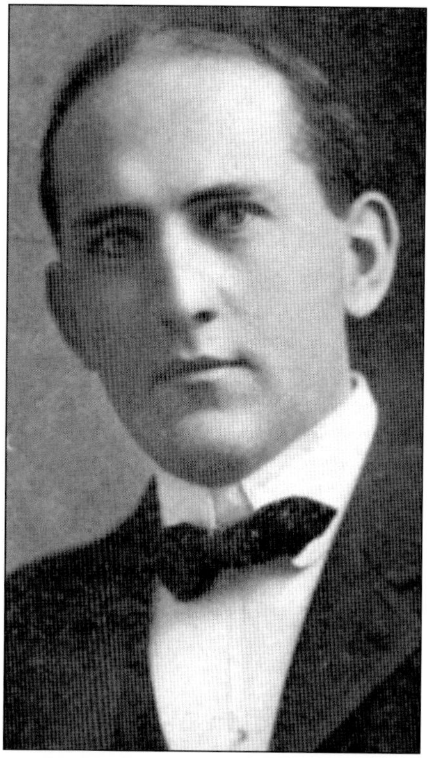

**Trace H. Paine, 1902.** From 1882 into the first two decades of the 1900s, Painesville native Trace Paine was a leading force in city firefighting. Joining Protection Fire shortly after moving to Ashtabula, Paine later served as a trustee and foreman. From 1902 to 1907, hardware retailer Paine served as part-time city fire chief, overseeing the operations of Ashtabula's eight volunteer fire companies as "one of the foremost in activity in the interest of the department." (Courtesy AFD Archive.)

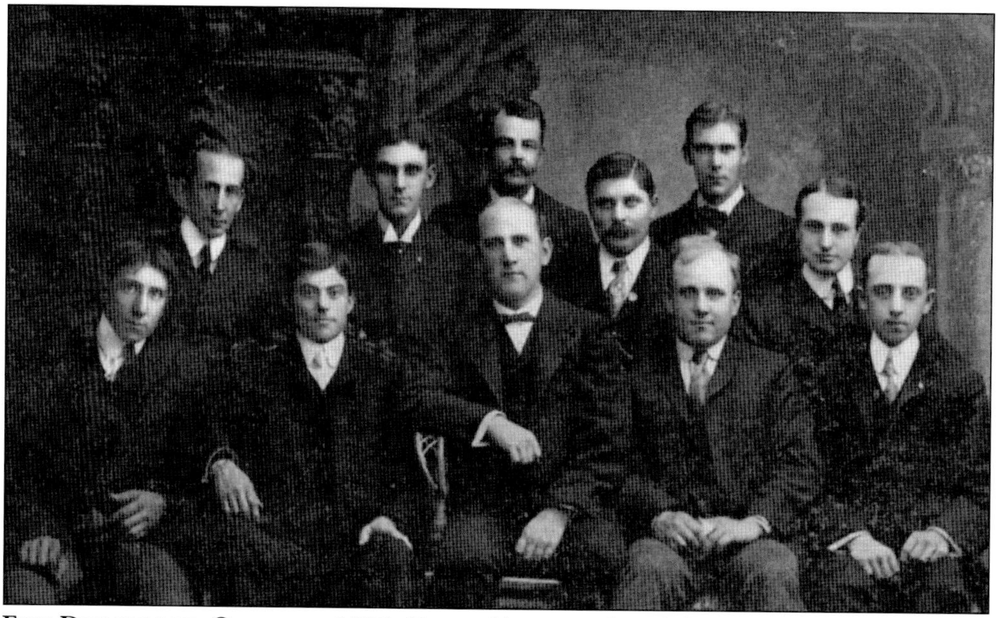

**Fire Department Officials, 1902.** Pictured here are, from left to right, (first row) Bonnar Hose foreman Henry Schrock, first assistant chief James T. Bowler, city fire chief Trace H. Paine, second assistant chief James Fitzgerald, and Protection Hose foreman A. T. Faulkner; (second row) Moore Hook and Ladder foreman A. J. Richardson, Neptune Hose foreman W. F. Butz, Nickel Plate Hose foreman Charles Flint, Humphrey Hose foreman E. G. Mapous, Raser Hose foreman J. T. Duffey, and Topky Hose foreman Clinton Watrous. (Courtesy AFD Archive.)

**EDWARD PIERCE HALL, 1902.** A leading Protection Fire volunteer was local attorney E. P. Hall (1864–1923), a Geneva native and graduate of Hudson's Western Reserve Academy. A principal in Ashtabula's Hoyt, Munsell and Hall law firm, Hall served Protection Fire as secretary in 1890 and treasurer in 1894–1895. A frequent member of Protection's executive committee, avid local historian Hall also served as Protection's department historian. (Courtesy AFD Archive.)

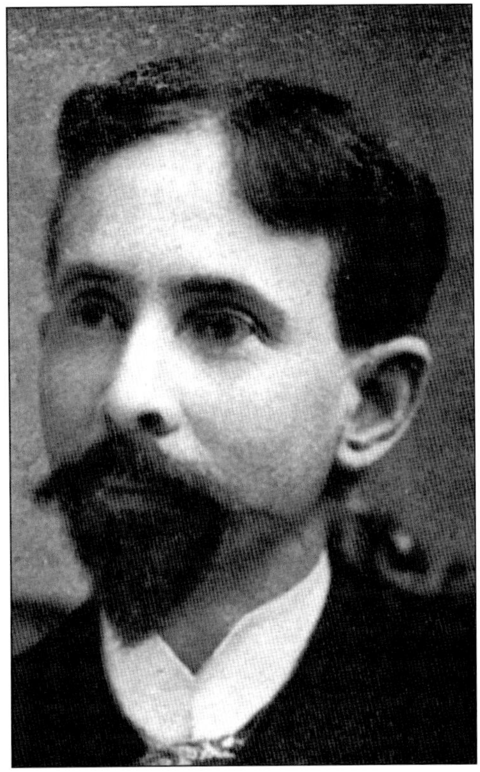

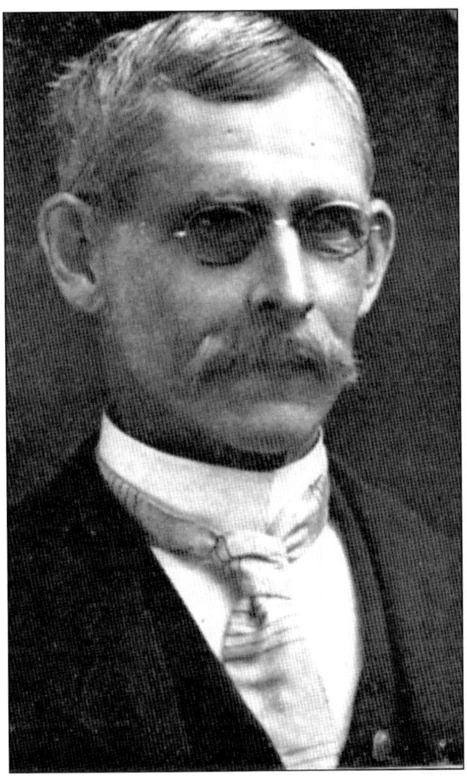

**W. F. SIEVERS, 1902.** German-born carpenter W. F. Sievers, one of Ashtabula's "most respected" men, was a prominent longtime Protection Fire volunteer. Civil War veteran Sievers joined Protection Fire after the war in 1865. Company foreman in 1894 and 1895, Sievers logged numerous terms as second assistant in the 1870s and 1880s, serving four terms as property manager in 1907–1909 and 1911. Sievers also served as assistant fire chief in the early 1890s. (Courtesy AFD Archive.)

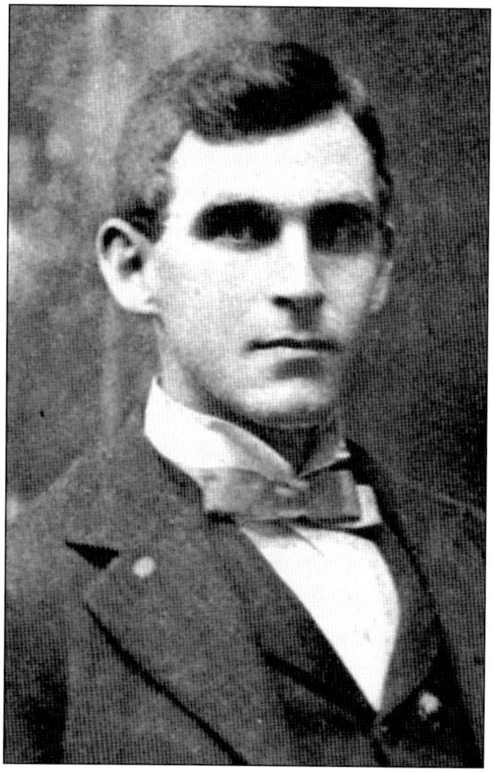

**WILLIAM F. BUTZ, 1902.** A longtime Neptune Hose volunteer firefighter and officer, leading Ashtabula saloon proprietor W. F. Butz was feted as "a pleasant, sociable man to meet and gentlemanly in his social and business matters." Butz served Neptune in several offices, including terms as foreman and second assistant foreman. Butz's tavern was famed as "a pleasant place to meet one's friends and throw off . . . the cares of life." (Courtesy AFD Archive.)

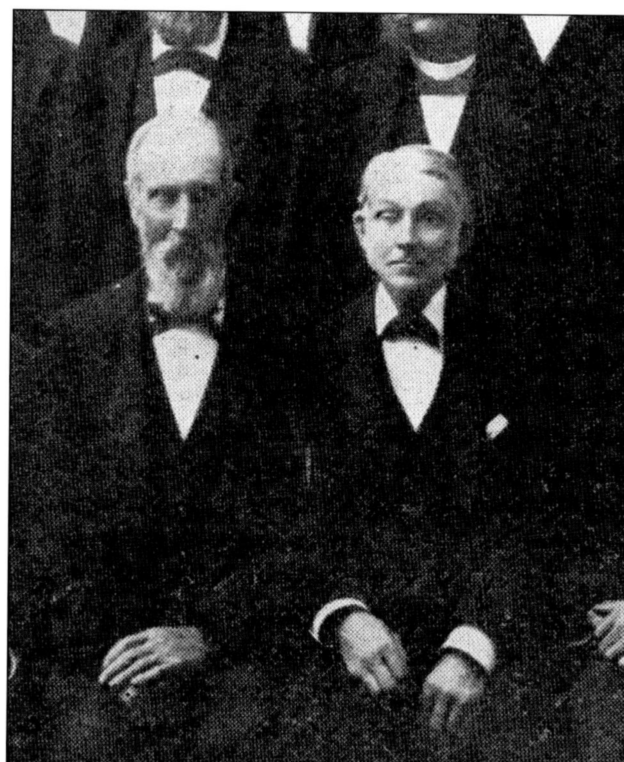

**G. A. KNAPP AND WILLIAM CLOVER, 1902.** Former Rising Sun Lodge No. 22 masters G. A. Knapp (left) and William Clover had strong ties to the Ashtabula Fire Department. Knapp, a longtime volunteer firefighter and former Protection Fire foreman, served a controversial term as fire chief during the infamous 1876 Ashtabula Bridge Disaster. Village mayor Clover, another longtime volunteer firefighter, served terms as foreman for Protection Fire and Lake Erie Hose. (Courtesy AFD Archive.)

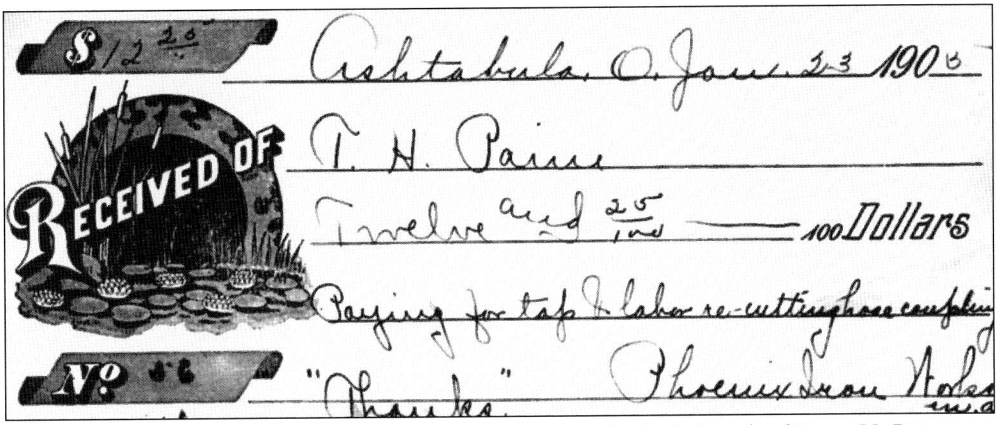

$12 25/100 Ashtabula, O. Jan. 23 1905

Received of T. H. Paine

Twelve and 25/100 Dollars

Paying for tap & labor re-cutting hose couplin

No. 86 "Thanks." Phoenix Iron Works

**PHOENIX IRON WORKS RECEIPT, JANUARY 23, 1905.** Ashtabula fire chief Trace H. Paine was the recipient of this $12.25 Phoenix Iron Works receipt for the recutting of fire hose couplings. Over a century later, the West Forty-eighth Street firm remains in business as Phoenix Machine and Foundry Company. (Courtesy AFD Archive.)

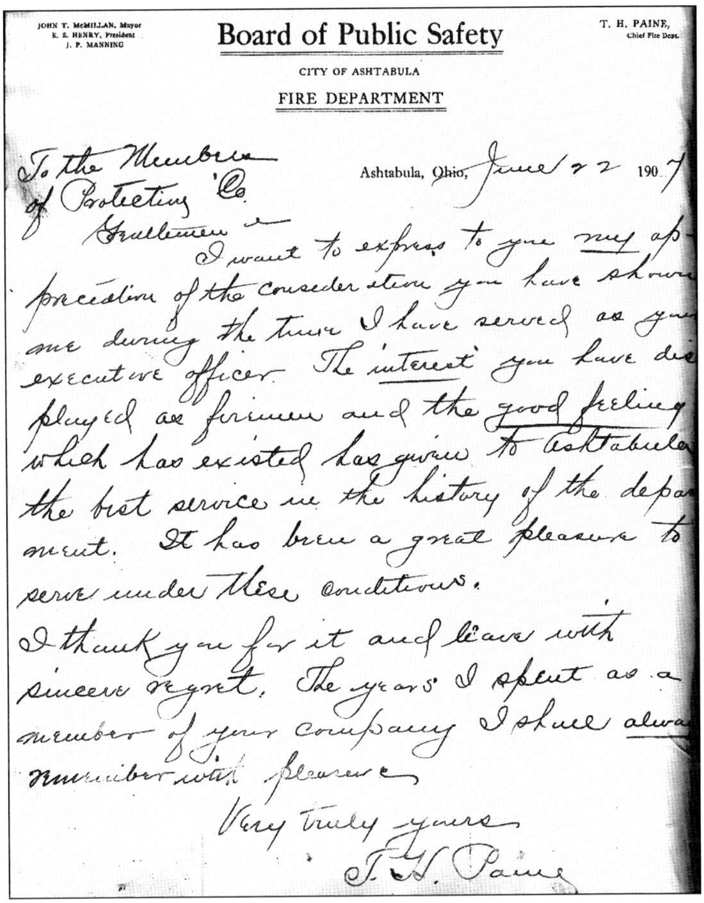

JOHN T. McMILLAN, Mayor
E. S. HENRY, President
J. P. MANNING

**Board of Public Safety**

T. H. PAINE,
Chief Fire Dept.

CITY OF ASHTABULA

FIRE DEPARTMENT

To the Members
of Protection Co.

Ashtabula, ~~Ohio,~~ June 22 1907

Gentlemen —

I want to express to you my appreciation of the consideration you have shown me during the time I have served as your executive officer. The interest you have displayed as firemen and the good feeling which has existed has given to Ashtabula the best service in the history of the department. It has been a great pleasure to serve under these conditions.

I thank you for it and leave with sincere regret. The years I spent as a member of your company I shall always remember with pleasure.

Very truly yours,
T. H. Paine

**TRACE H. PAINE LETTER, JUNE 1907.** A Protection Fire volunteer since 1882 and the city's part-time fire chief since 1902, prominent downtown hardware retailer Paine retired as chief in June 1907. Wrote Paine to Protection Fire volunteers, "The interest you have displayed as firemen and the good feeling which has existed has given to Ashtabula the best service in the history of the department." (Courtesy AFD Archive.)

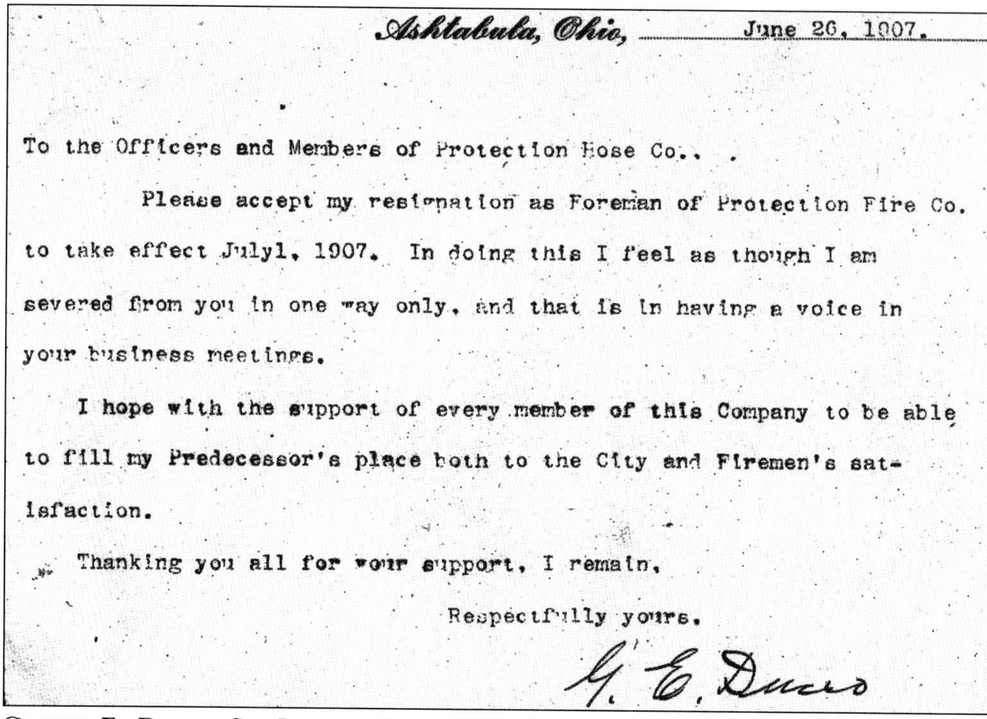

Ashtabula, Ohio, June 26, 1907.

To the Officers and Members of Protection Hose Co.,

Please accept my resignation as Foreman of Protection Fire Co. to take effect July1, 1907. In doing this I feel as though I am severed from you in one way only, and that is in having a voice in your business meetings.

I hope with the support of every member of this Company to be able to fill my Predecessor's place both to the City and Firemen's satisfaction.

Thanking you all for your support, I remain,

Respectfully yours,

G. E. Ducro

**George E. Ducro Sr. Letter, June 1907.** Appointed to succeed part-time Ashtabula Fire Department chief Trace Paine, veteran Protection Fire volunteer and leading Ashtabula furniture retailer George E. Ducro Sr. wrote to Protection's officers and members, tendering his resignation as company foreman: "I hope with the support of every member of this Company to be able to fill my Predecessor's place both to the City and Fireman's satisfaction." (Courtesy AFD Archive.)

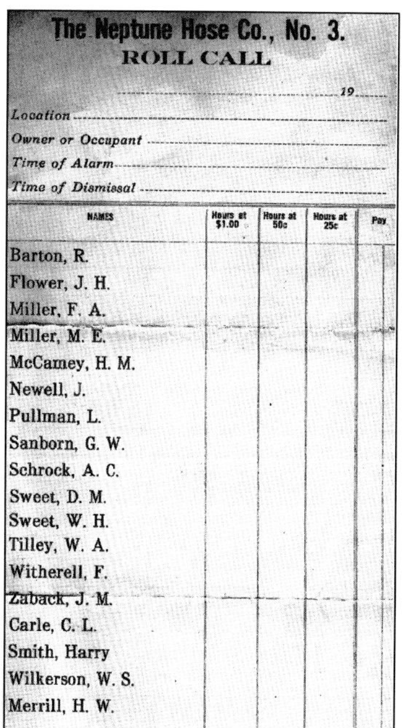

The Neptune Hose Co., No. 3.
ROLL CALL

.......... 19.....

Location ..........
Owner or Occupant ..........
Time of Alarm ..........
Time of Dismissal ..........

| NAMES | Hours at $1.00 | Hours at 50c | Hours at 25c | Pay |
|---|---|---|---|---|
| Barton, R. | | | | |
| Flower, J. H. | | | | |
| Miller, F. A. | | | | |
| Miller, M. E. | | | | |
| McCamey, H. M. | | | | |
| Newell, J. | | | | |
| Pullman, L. | | | | |
| Sanborn, G. W. | | | | |
| Schrock, A. C. | | | | |
| Sweet, D. M. | | | | |
| Sweet, W. H. | | | | |
| Tilley, W. A. | | | | |
| Witherell, F. | | | | |
| ~~Zaback, J. M.~~ | | | | |
| Carle, C. L. | | | | |
| Smith, Harry | | | | |
| Wilkerson, W. S. | | | | |
| Merrill, H. W. | | | | |

**Neptune Hose Company Roll Call, 1907–1912.** The Neptune Hose roster included R. Barton, J. H. Flower, F. A. Miller, M. E. Miller, H. M. McCamey, J. Newell, L. Pullman, G. W. Sanborn, A. C. Schrock, D. M. Sweet, W. H. Sweet, W. A. Tilley, F. Witherell, J. M. Zaback, C. L. Carle, Harry Smith, W. S. Wilkerson, and H. W. Merrill. Volunteers were remunerated for services rendered, $1 for the first hour, and 50¢ an hour thereafter. (Courtesy AFD Archive.)

*VICK'S STANDARD DIRECTORY,* **1907.** In its volunteer days, the Ashtabula Fire Department operated eight volunteer fire companies of 18 men and three executive officers. The companies were Protection Fire, Moore Hook and Ladder, Neptune Hose, the Harbor's Topky Hose, Bonnar Hose, Raser Hose, Nickel Plate Hose, and Humphrey Hose. Department leadership in early 1907 included fire chief Trace Paine, assistant chief James Bowler, and second assistant chief James Fitzgerald. (Courtesy Ashtabula County District Library.)

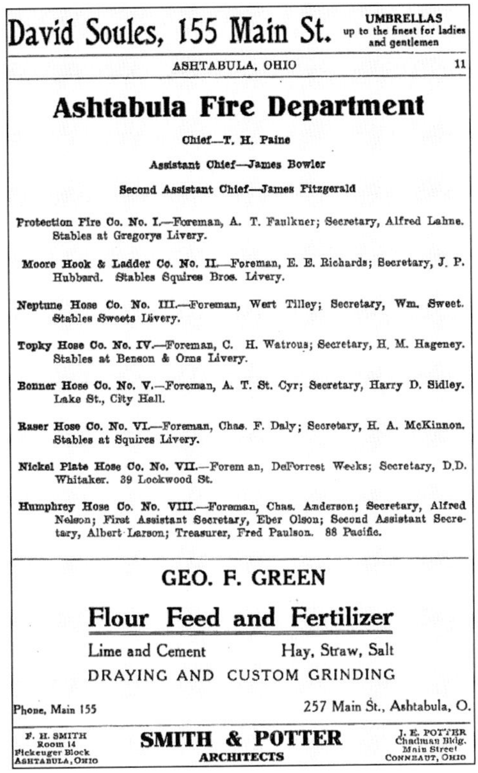



ASHTABULA, OHIO 11

**Ashtabula Fire Department**

Chief—T. H. Paine

Assistant Chief—James Bowler

Second Assistant Chief—James Fitzgerald

Protection Fire Co. No. I.—Foreman, A. T. Faulkner; Secretary, Alfred Lahne. Stables at Gregorys Livery.

Moore Hook & Ladder Co. No. II.—Foreman, E. E. Richards; Secretary, J. P. Hubbard. Stables Squires Bros. Livery.

Neptune Hose Co. No. III.—Foreman, Wert Tilley; Secretary, Wm. Sweet. Stables Sweets Livery.

Topky Hose Co. No. IV.—Foreman, C. H. Watrous; Secretary, H. M. Hageney. Stables at Benson & Orns Livery.

Bonner Hose Co. No. V.—Foreman, A. T. St. Cyr; Secretary, Harry D. Sidley. Lake St., City Hall.

Raser Hose Co. No. VI.—Foreman, Chas. F. Daly; Secretary, H. A. McKinnon. Stables at Squires Livery.

Nickel Plate Hose Co. No. VII.—Forem an, DeForrest Weeks; Secretary, D.D. Whitaker. 39 Lockwood St.

Humphrey Hose Co. No. VIII.—Foreman, Chas. Anderson; Secretary, Alfred Nelson; First Assistant Secretary, Eber Olson; Second Assistant Secretary, Albert Larson; Treasurer, Fred Paulson. 88 Pacific.



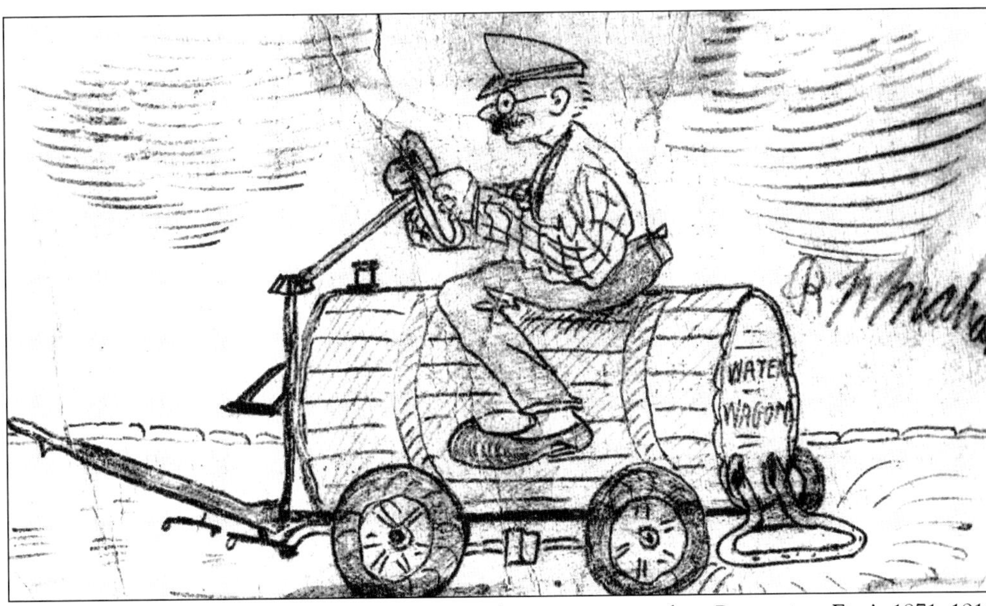

**RAY N. MAHAN, 1910.** Among the historical items preserved in Protection Fire's 1871–1918 journal are cartoons created by future 1929–1937 city fire chief Ray N. Mahan, a 15-year volunteer Protection firefighter and frequent company officer (property manager, 1904–1906; first assistant foreman, 1907–1910). The mustachioed Mahan, seen here in a self-caricature atop a Protection Fire "water wagon," created numerous drawings poking good-natured fun at Protection Fire and its volunteers. (Courtesy AFD Archive.)

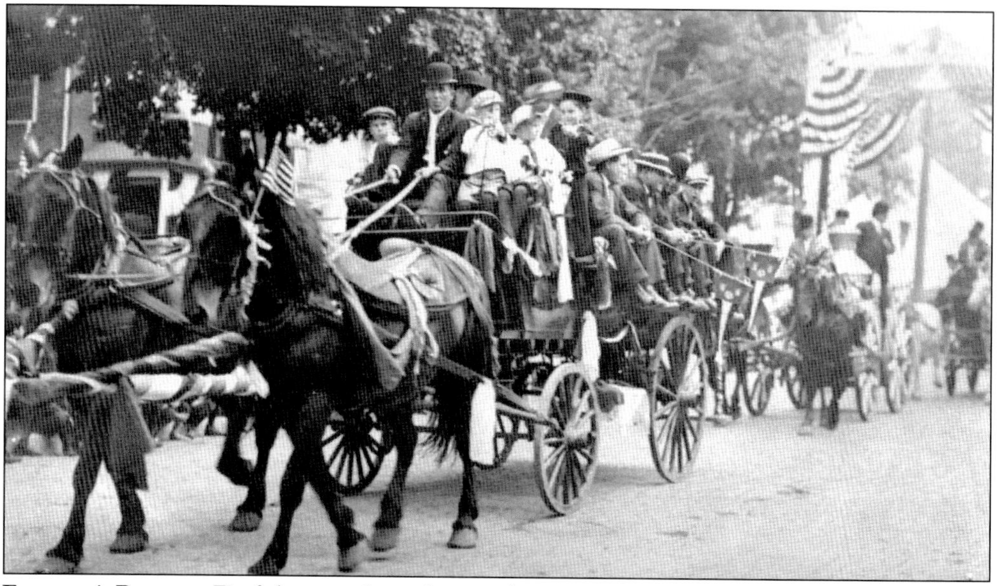

**FIREMEN'S PARADE.** Firefighters and parades are a longtime mainstay, as evidenced by Ashtabula's September 6, 1911, gala "Firemen's Parade." Feted by the Ashtabula *Beacon-Record* as "wholly satisfactory," the afternoon parade was so well attended that downtown sidewalks were said to be "black with populace, which filled every available window and doorway along the route." The Firemen's Parade included fire chief George E. Ducro Sr., city safety director Dr. I. M. Hadcock, and Ashtabula's volunteer Raser (below), Nickel Plate, Topky, Neptune, Moore Hook and Ladder, and Protection companies. Visiting participants included Ashtabula Township's Fargo and East Side Chemical companies, the Conneaut Fire Department, the Painesville Fire Department band, and fire chiefs from Youngstown, Canton, Warren, and Conneaut. Following the parade, Ashtabula firefighters hosted a demonstration of the city's newly delivered 1911 Robinson Jumbo auto fire truck at Lake and Main Avenues. (Courtesy AFD Archive.)

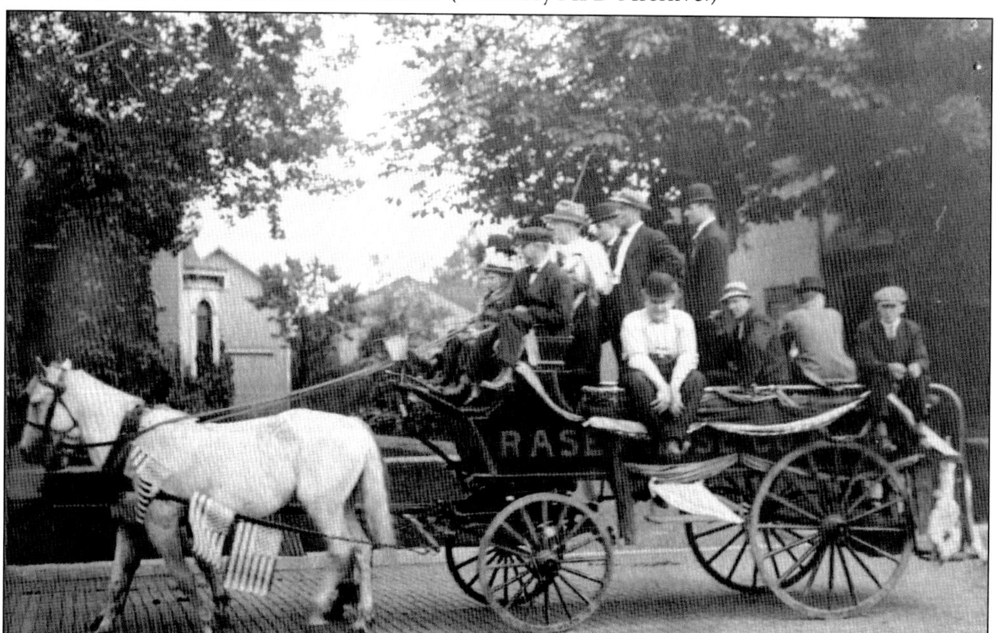

**PROTECTION FIRE ROLL CALL.** With the February 1912 opening of the paid fire department's Uptown Fire Station, six of Ashtabula's eight volunteer fire companies disbanded, including 62-year-old Protection Fire. Protection's last roster was as follows: A. D. Amsden, C. R. Canfield, E. G. Dickenson, A. T. Faulkner, A .L. Gregory, Carl Herrick, Alfred Kahne, E. R. Kelly, Ray Mahan, J. H. McGiffert, Fred and William Morrison, W. F. Sievers, Don Reed, Charles Tyler, A. T. Ullman, Frank Wagner, and J. L. Wilson. (Courtesy AFD Archive.)

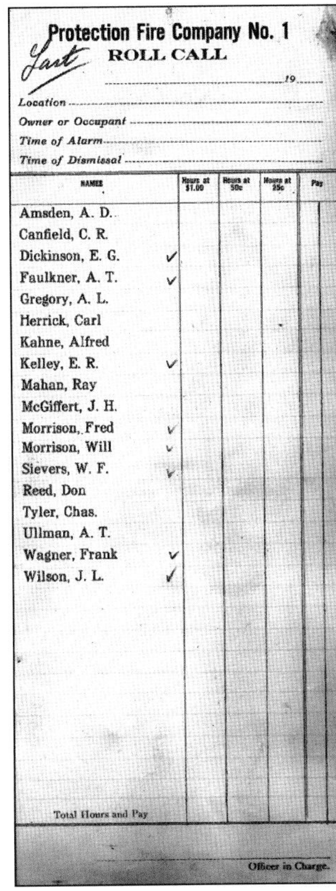
Protection Fire Company No. 1

Last ROLL CALL

19

Location

Owner or Occupant

Time of Alarm

Time of Dismissal

| NAMES | Hours at $1.00 | Hours at 50c | Hours at 25c | Pay |
|---|---|---|---|---|
| Amsden, A. D. | | | | |
| Canfield, C. R. | | | | |
| Dickinson, E. G. | ✓ | | | |
| Faulkner, A. T. | ✓ | | | |
| Gregory, A. L. | | | | |
| Herrick, Carl | | | | |
| Kahne, Alfred | | | | |
| Kelley, E. R. | ✓ | | | |
| Mahan, Ray | | | | |
| McGiffert, J. H. | | | | |
| Morrison, Fred | ✓ | | | |
| Morrison, Will | ✓ | | | |
| Sievers, W. F. | ✓ | | | |
| Reed, Don | | | | |
| Tyler, Chas. | | | | |
| Ullman, A. T. | | | | |
| Wagner, Frank | ✓ | | | |
| Wilson, J. L. | ✓ | | | |
| Total Hours and Pay | | | | |

Officer in Charge.

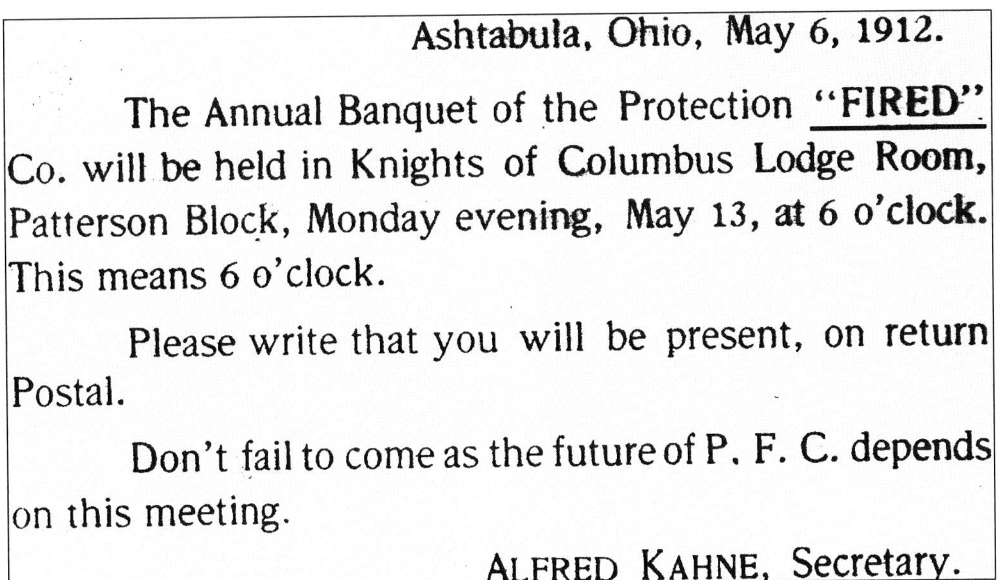
Ashtabula, Ohio, May 6, 1912.

The Annual Banquet of the Protection "FIRED" Co. will be held in Knights of Columbus Lodge Room, Patterson Block, Monday evening, May 13, at 6 o'clock. This means 6 o'clock.

Please write that you will be present, on return Postal.

Don't fail to come as the future of P. F. C. depends on this meeting.

ALFRED KAHNE, Secretary.

**PROTECTION "FIRED" COMPANY.** Retired in February 1912 by the paid professional fire department and its new headquarters Uptown Fire Station on Park Avenue, the self-designated Protection *Fired* Company considered its future at its May 13, 1912, annual meeting, where "memories of the past were recalled . . . and the stories flowed fast and furious." Members voted overwhelmingly to continue "for old time's sake" as a purely social organization. (Courtesy AFD Archive.)

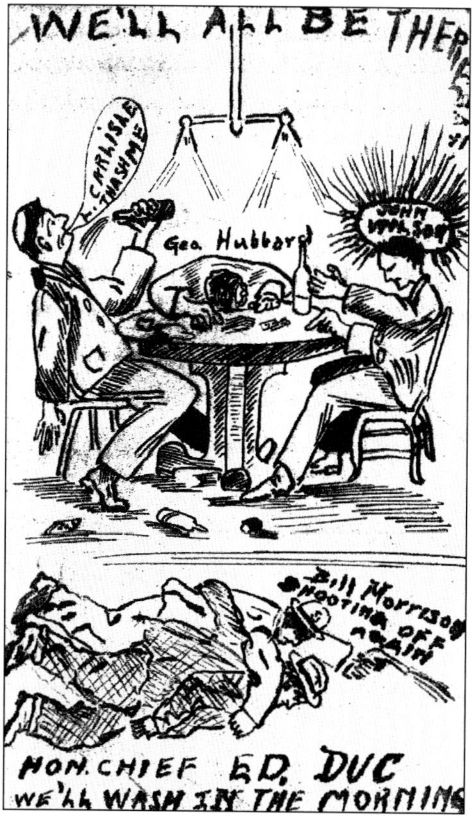


**PROTECTION FIRE COMPANY ANNUAL PICNIC REPLY POSTAL, 1912.** Good-natured ribbing has long been a part of the firefighting brotherhood, as evidenced by Protection Fire veteran Ray Mahan's cartoon lampoonings of the fun-loving, hard-drinking foibles of his fellow Protection brothers, bandmaster L. T. Carlisle, retiree George C. Hubbard, historian John L. Wilson, executive committee member William H. (Bill) Morrison, and Protection alumni and city fire chief George E. Ducro Sr. (Courtesy AFD Archive.)

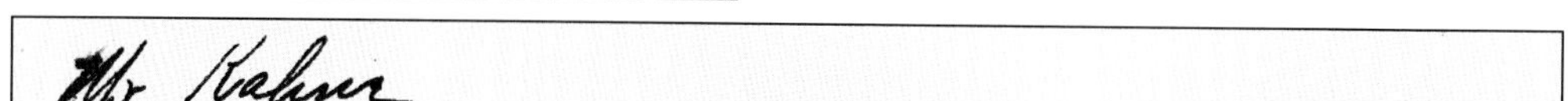
Mr Kahn.
Dear Sir,
I got the card you expected my husband to get, inviting him to your drunk. He has taken those in for the past twenty years, and every time come home sick and been sick for a week afterward. I dont want you to send him any more for I shall not let him attend. It made me so mad when I got this card that if I'd had you here I'd of wrung your neck. Never send him another.

**PROTECTION FIRE COMPANY ANNUAL PICNIC REPLY POSTAL, 1913.** Protection Fire's annual picnics were fun-loving socials, known for their spirited card games and free-flowing beer and wine. This Protection wife did not mince words: "I got the card you expected my husband to get, inviting him to your drunk. It made me so mad when I got this card that if I'd had you here I'd have wrung your neck. Never send him another." (Courtesy AFD Archive.)

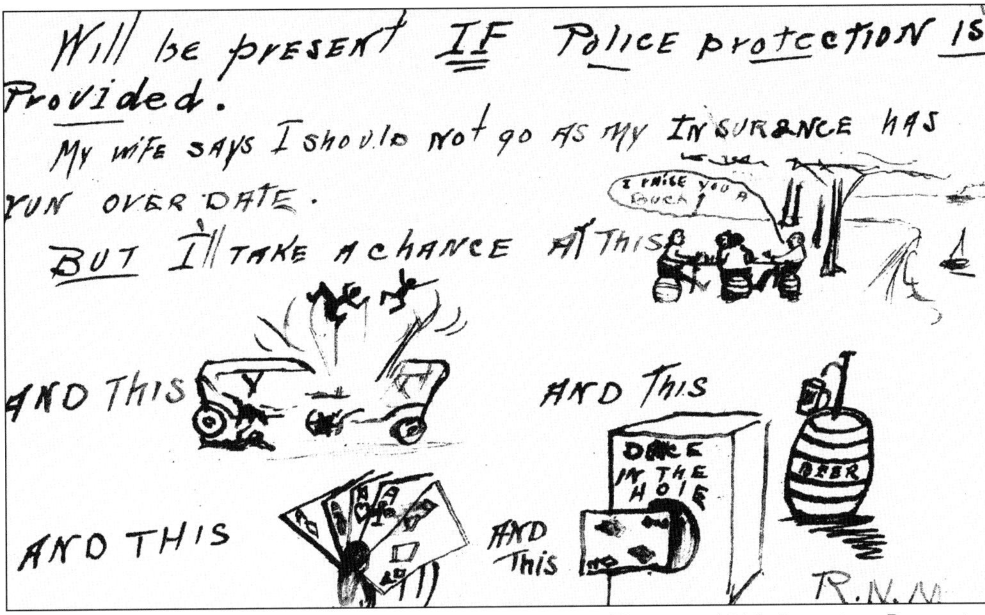


**Protection Fire Company Annual Picnic Return Postal, 1916.** Longtime Protection Fire volunteer and future city fire chief Ray N. Mahan created this humorous RSVP, poking fun at the ebullient card playing and beer drinking that typified Protection Fire annual picnics at fire chief George E. Ducro's lakefront Lake Road East farm in Ashtabula Township. (Courtesy AFD Archive.)

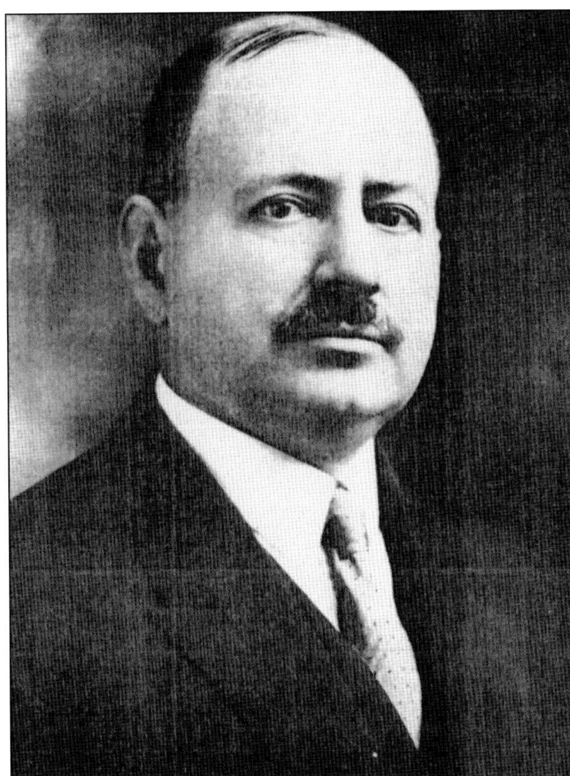

**Alfred Kahne.** Owner of downtown's popular Central News Room, German-born Alfred Kahne was a longtime Protection Fire officer. Said the *Star Beacon* of Protection Fire veterans Kahne and A. L. Gregory in 1950, "These old-timers willingly concede that the present Division of Fire is more efficient than its predecessors, but they claim it lacks the color and glamour rampant in the era of horse-drawn engines and uniformed drill teams." (Courtesy Harbor-Topky Memorial Library.)

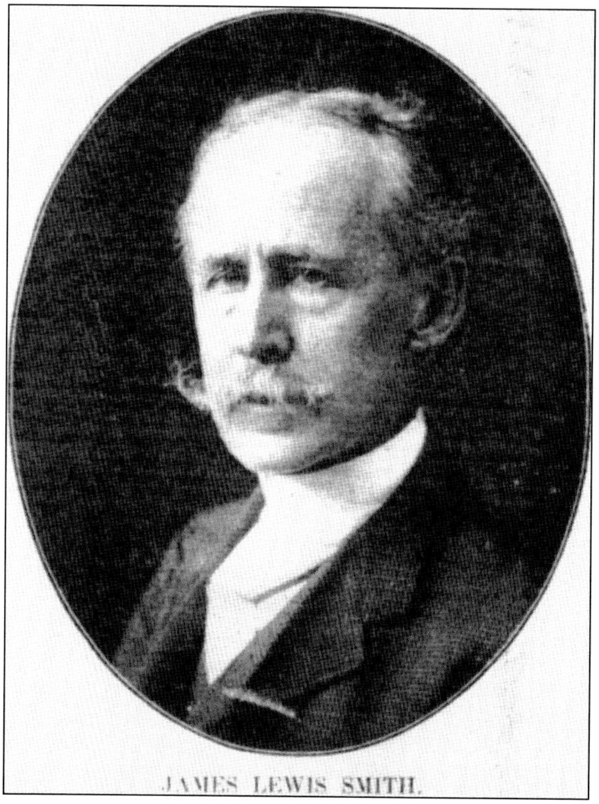


**JAMES LEWIS SMITH.** As the aristocrat of city volunteer fire companies, Protection Fire's roster encompassed many of Ashtabula's leading business and civic leaders, including businessman and philanthropist James L. Smith (1850–1919). Smith spearheaded the c. 1895 Ashtabula Bridge Disaster memorial at Chestnut Grove Cemetery and donated 4.61 acres for the construction of Ashtabula General Hospital. Smith's bequest funded creation of the James L. Smith Home for Aged Women, today's Park Haven. (Courtesy Harbor-Topky Memorial Library.)

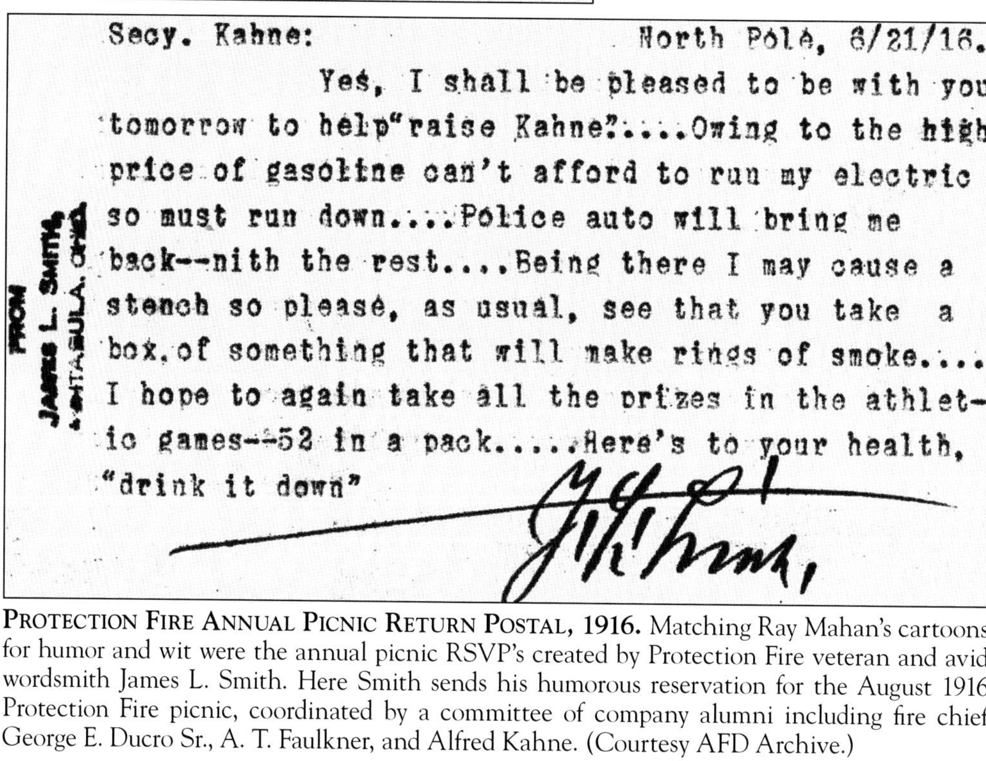
FROM JAMES L. SMITH, ASHTABULA, OHIO

Secy. Kahne: North Pole, 8/21/16.

Yes, I shall be pleased to be with you tomorrow to help"raise Kahne"....Owing to the high price of gasoline can't afford to run my electric so must run down....Police auto will bring me back--with the rest....Being there I may cause a stench so please, as usual, see that you take a box of something that will make rings of smoke.... I hope to again take all the prizes in the athletic games--52 in a pack....Here's to your health, "drink it down"

J L Smith,

**PROTECTION FIRE ANNUAL PICNIC RETURN POSTAL, 1916.** Matching Ray Mahan's cartoons for humor and wit were the annual picnic RSVP's created by Protection Fire veteran and avid wordsmith James L. Smith. Here Smith sends his humorous reservation for the August 1916 Protection Fire picnic, coordinated by a committee of company alumni including fire chief George E. Ducro Sr., A. T. Faulkner, and Alfred Kahne. (Courtesy AFD Archive.)

# *Two*

# The Ashtabula Bridge Disaster

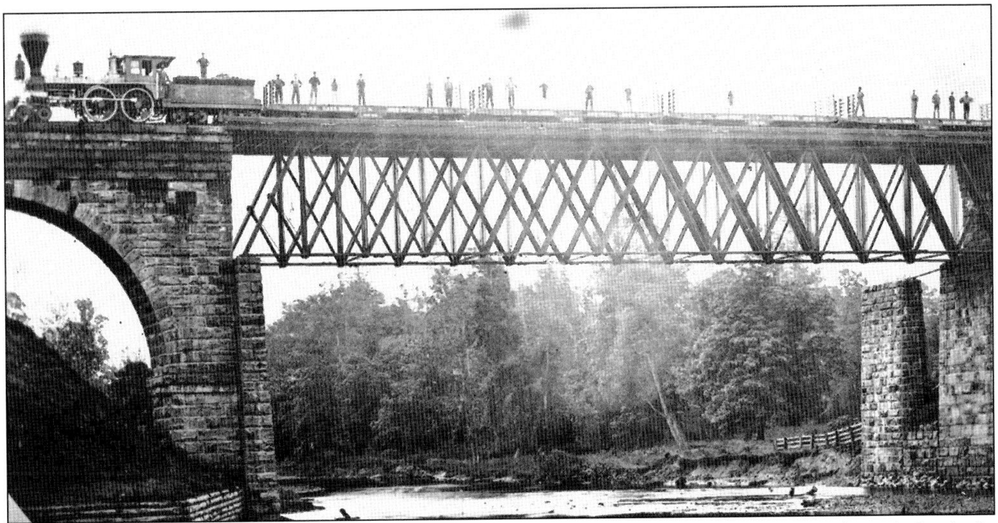

**Completion of the Bridge, 1865.** A source of pride for growing 2,500-resident Ashtabula and the Lake Shore and Michigan Southern Railway was this experimental 150-foot double track steel Howe Truss–styled bridge spanning the Ashtabula River gulf. Built in 1864–1865, the 11-year-old bridge collapsed under the railroad's "pride of the line" *Pacific Express* during a blinding three-foot blizzard on December 29, 1876. Just 600 yards short of the Ashtabula depot, the 159-passenger *Pacific Express* plunged nearly 80 feet into the gulf at 7:28 p.m. Onboard kerosene heating stoves touched off an inferno that quickly consumed the wreckage. The disaster claimed 92 lives, with 64 injured passengers and crew surviving the crash and ensuing blaze. Volunteer firefighters from Lake Erie Hose, Protection Fire, and Neptune Hose worked feverishly alongside citizens to rescue disaster victims. Noted the Ashtabula *Sentinel*, "There were noble hearts, and manly hands, and many were the heroic deeds done at that fearful time. The wonder is not that so few were saved, but that so many escaped." (Courtesy Ashtabula Star Beacon.)

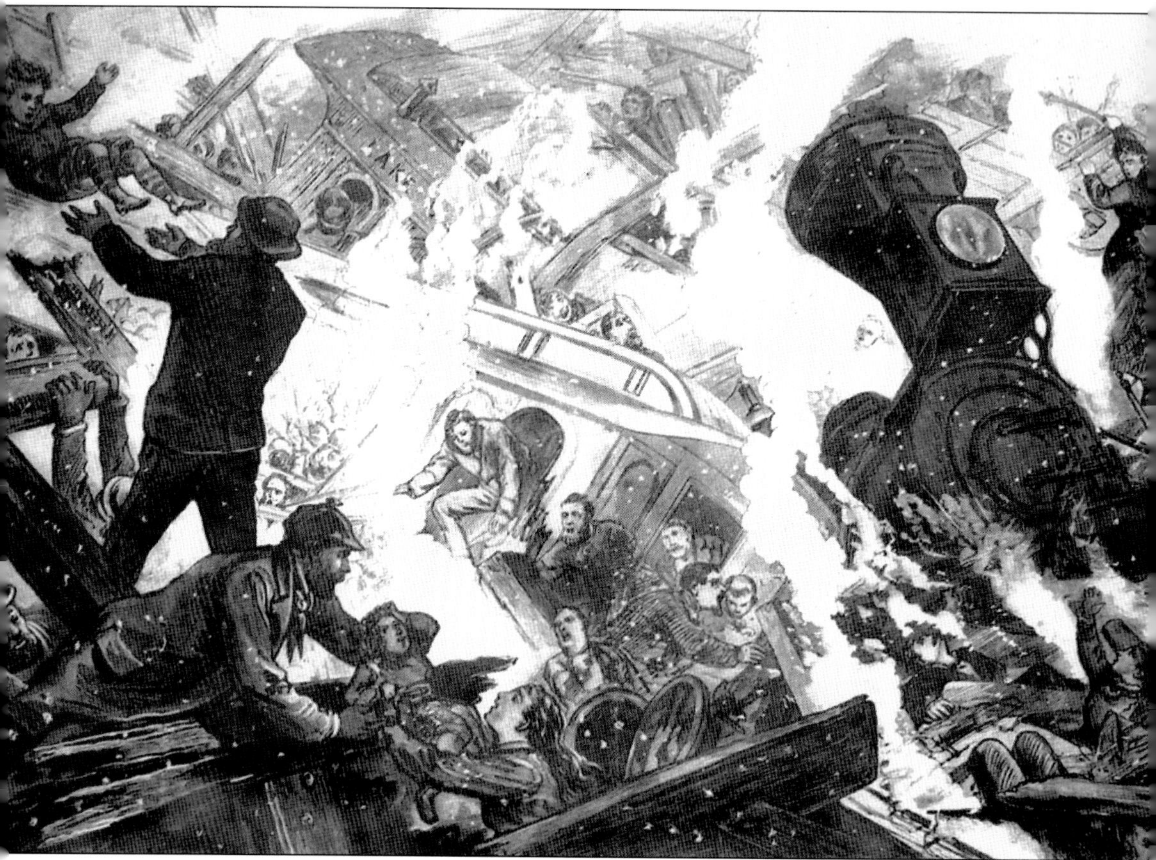

**"The Horrible Railroad Disaster in Ohio, December 29th—Rescuing The Passengers From The Frightful Wreck—Heroic Efforts To Save Women and Children From The Valley of Death," 1876.** This Cleveland *Plain Dealer* engraving depicts Ashtabula Bridge Disaster rescue efforts. A quickly assembled coroner's jury issued a still-controversial stinging rebuke of Ashtabula's volunteer firefighters, noting that they "seemed to have been so overwhelmed by the fearful calamity that they lost all presence of mind," citing their failure to throw water on the growing fire. Village fire chief G. A. Knapp, described in news accounts as "slow and lymphatic in temperament, addicted to constant use of intoxicating liquors, and in every way unfit for so trying an emergency," was chastised by the jury for failing to order firefighters to extinguish the initial small fires, which quickly turned the wreckage into an inferno that burned scores of crash survivors alive. Local firefighters issued a rebuttal to the jury report, insisting firefighting attempts would have been useless and their efforts had been better used on immediate search and rescue operations. (Courtesy Cleveland Plain Dealer.)

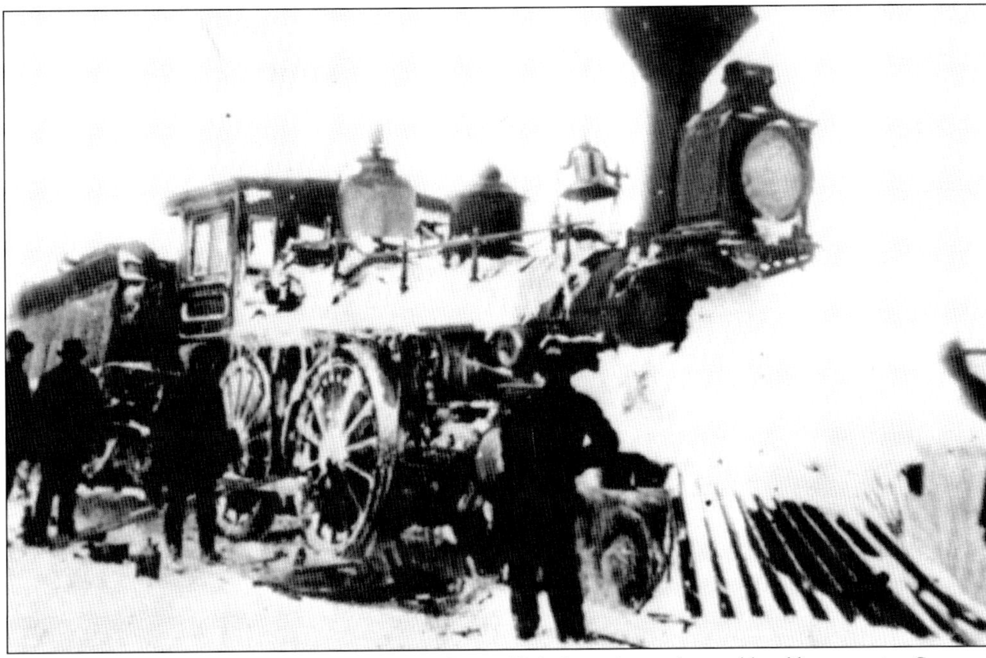

**SOCRATES, DECEMBER 30, 1876.** Engineer Daniel McGuire's westbound lead locomotive *Socrates* was the surviving remnant of the Lake Shore and Michigan Southern's ill-fated *Pacific Express.* Shown here stopped "100 feet west from the chasm" created by the deadly bridge collapse, the *Socrates*' coal tender (left) derailed and was thrown against the locomotive as the bridge deck collapsed, a vivid illustration of the *Socrates*' desperate struggle to clear the west abutment. (Courtesy Ashtabula Star Beacon.)

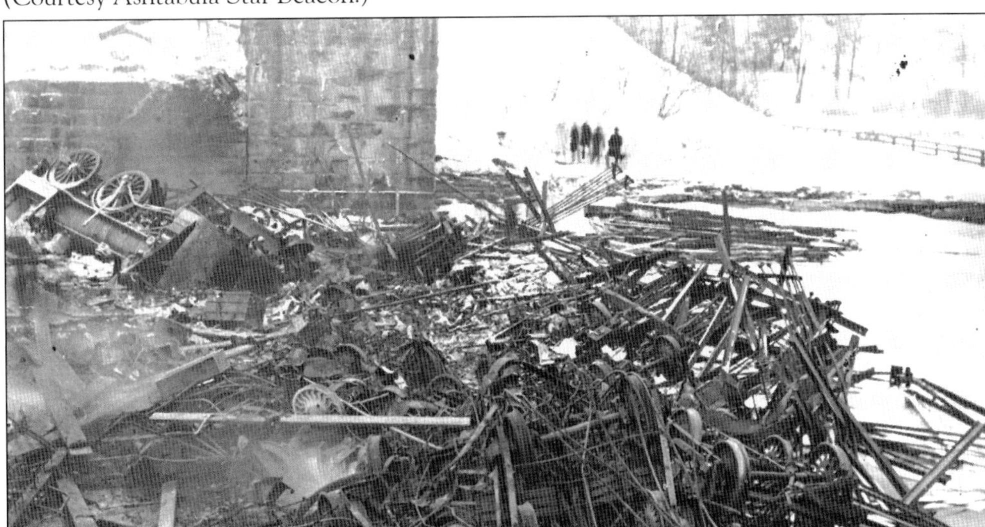

**ASHTABULA BRIDGE DISASTER WRECKAGE, DECEMBER 30, 1876.** This famed morning-after photograph was taken by professional Ashtabula photographer Fred W. Blakeslee, a longtime Protection Fire volunteer and eight-year village fire chief. Blakeslee's memorable photograph was reproduced in the early 1900s on color penny postcards recalling the historic disaster. The last disaster survivor, Ohio native Effie Neeley, died at age 101 in March 1960. (Photograph Fred W. Blakeslee; courtesy Ashtabula Star Beacon.)

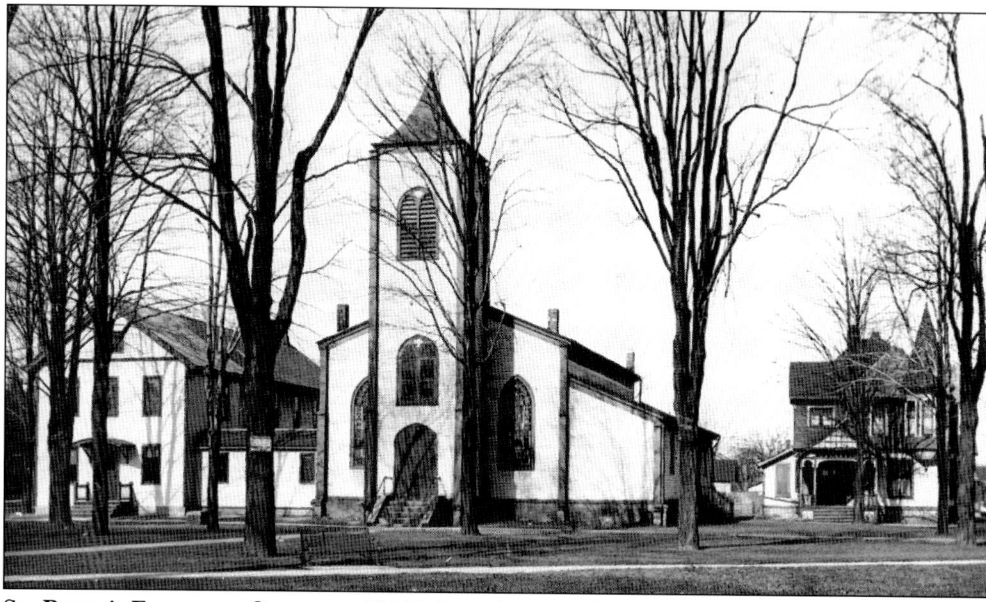

**St. Peter's Episcopal Church, 4901 Main Avenue.** First Methodist Church and St. Peter's Episcopal Church (pictured) hosted standing-room-only funeral services for the Ashtabula Bridge Disaster's dead. An ecumenical service was held at First Methodist. The service at St. Peter's, conducted by rector Rev. James Moore and Cleveland's Rev. George Carter, hit especially close to home. One of the victims was Carter's brother-in-law, Rev. A. H. Washburn, beloved rector of Grace Episcopal Church, Cleveland. (Courtesy St. Peter's Episcopal Church.)

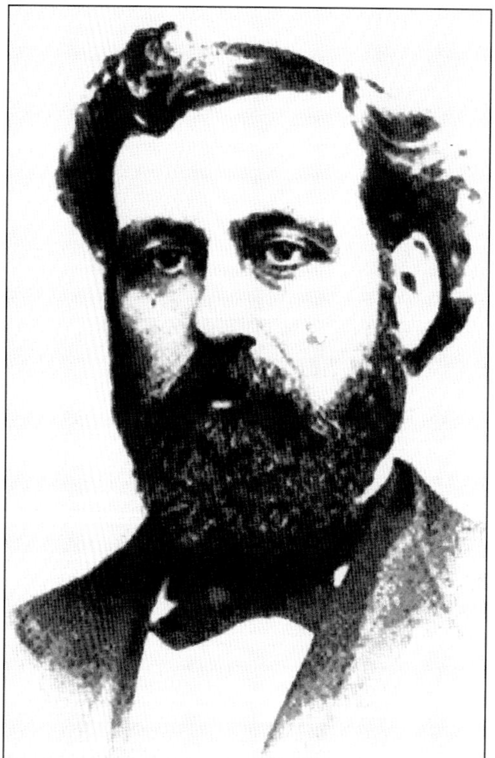

**Philip Paul Bliss.** With many Ashtabula Bridge Disaster victims mangled or burned beyond recognition, unidentified victims from Iowa, Illinois, Wisconsin, Ohio, New York, and Massachusetts were buried in a mass grave at Chestnut Grove Cemetery. Among the unidentified dead was famed American hymn writer and composer Philip P. Bliss, feted for hymns like *Hallelujah! What a Savior* and the musical score for *It Is Well With My Soul.* (Author's collection.)

**CORONER'S JURY, 1876–1877.** The day after the horrific Ashtabula Bridge Disaster, a local coroner's jury was assembled to investigate the incident, an inquest that spanned 68 days. Community leaders chosen for the coroner's jury included T. D. Faulkner, Edward G. Pierce, future Protection Fire foreman George W. Dickinson, Henry H. Perry, F. A. Pettibone, and prominent retailer Henry Lawton Morrison (right), a four-term mayor, councilman, school board member, county commissioner, and charter board member of the Pittsburgh, Youngstown and Ashtabula Railway. Justice of the peace Edward W. Richards served as acting coroner. Longtime attorney Theodore Hall (below), appointed a common pleas judge in 1902, was selected as the jury's legal counsel. The coroner's jury ruled the Lake Shore and Michigan Southern Railway was fully liable for the accident, citing poor bridge design, a lack of competent bridge inspection, and the railway's failure to provide self-extinguishing heating apparatus on the *Pacific Express* as required by law. (Courtesy AFD Archive.)

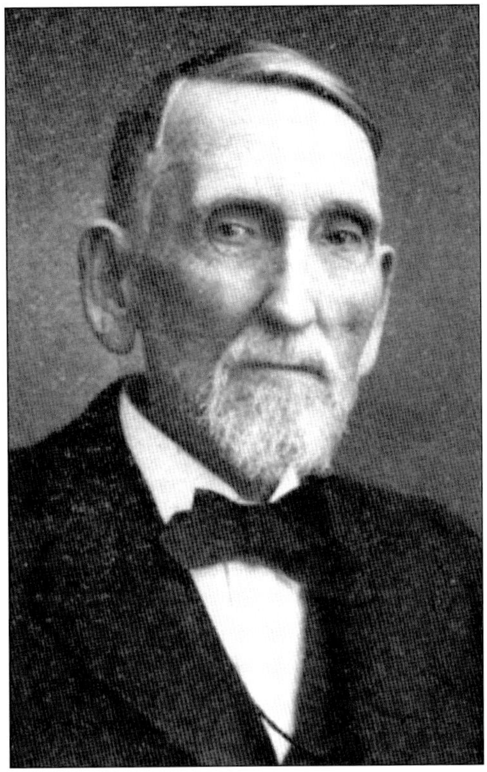

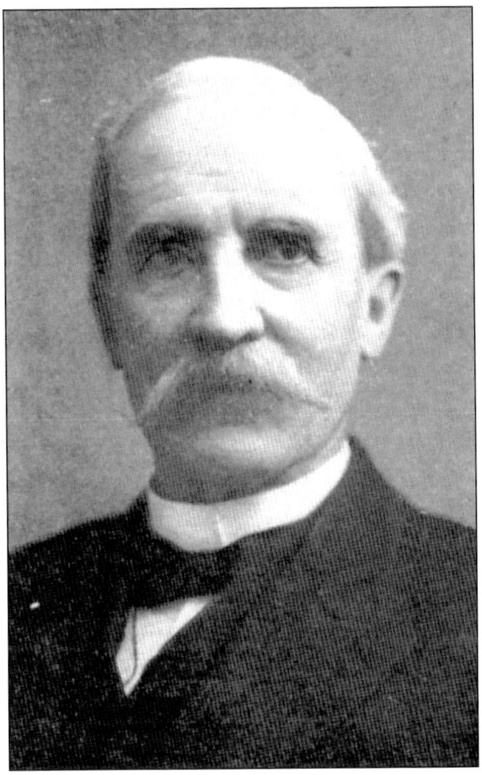

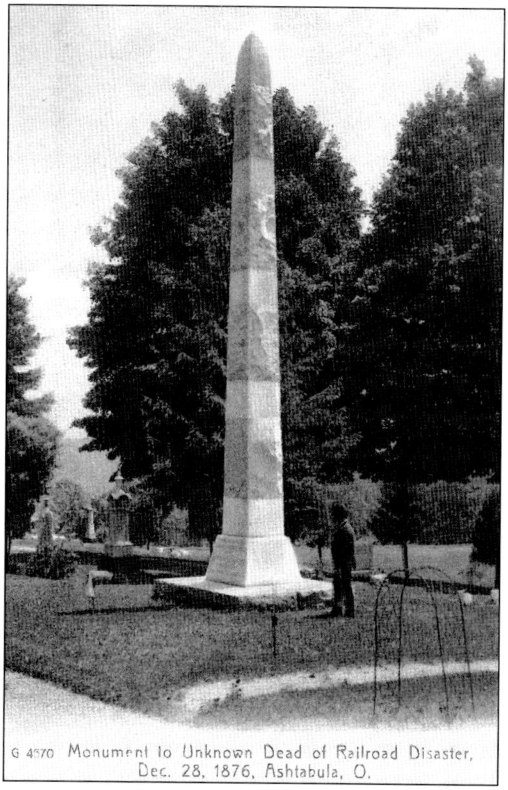


**MONUMENT TO UNKNOWN DEAD OF RAILROAD DISASTER.** Thanks to the efforts of Ashtabula's Knights of Pythias Unity Lodge No. 133, a 35-foot granite obelisk was erected at Chestnut Grove Cemetery in memory of the Ashtabula Bridge Disaster's unidentified dead. Unidentified local victims included Garwood B. Stow of Geneva and Albert H. Stockwell and George W. Kepler of Ashtabula. Dedicated before a crowd of 5,000 on Memorial Day 1895, donors included United States president William McKinley. (Author's collection.)

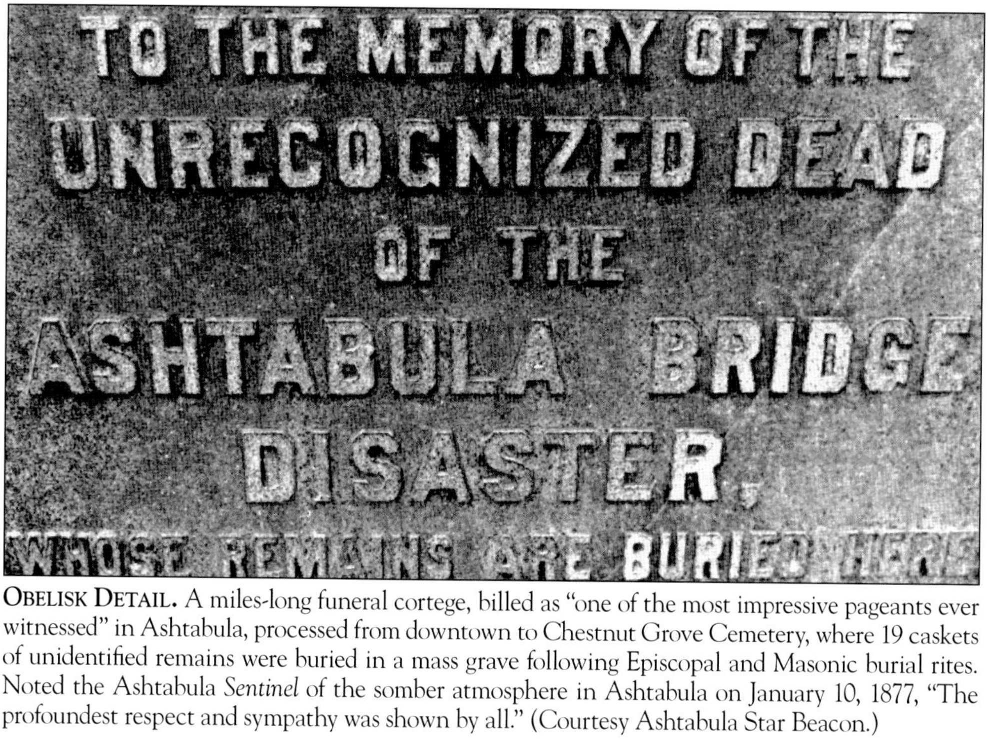


**OBELISK DETAIL.** A miles-long funeral cortege, billed as "one of the most impressive pageants ever witnessed" in Ashtabula, processed from downtown to Chestnut Grove Cemetery, where 19 caskets of unidentified remains were buried in a mass grave following Episcopal and Masonic burial rites. Noted the Ashtabula *Sentinel* of the somber atmosphere in Ashtabula on January 10, 1877, "The profoundest respect and sympathy was shown by all." (Courtesy Ashtabula Star Beacon.)

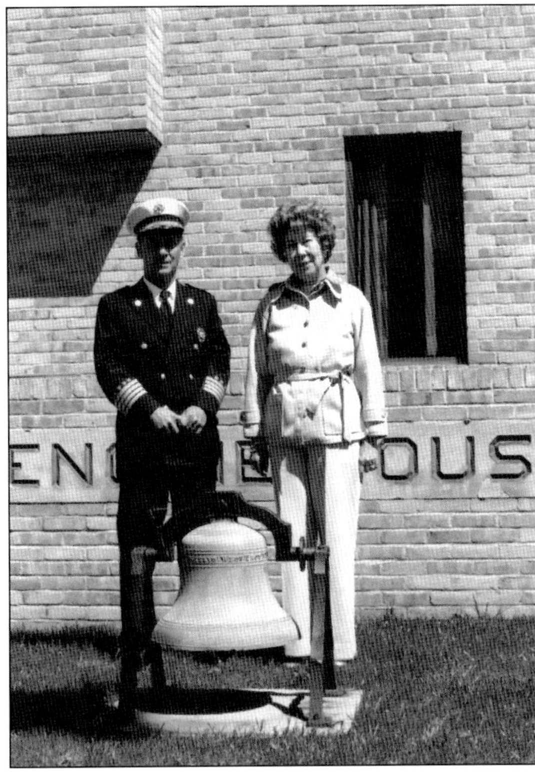

**Ashtabula Memorial Bell, November 1975.** Ashtabula fire chief Charles L. Mosier and Mary Poole Crawford admire the department's recovered Lake Erie Hose fire bell. The 200-pound bronze bell, cast in 1854 by Meneeley and Company (1826–1952), called Lake Erie Hose volunteers to the Ashtabula Bridge Disaster. Crawford's aunt, Missouri resident Lena Fox Schlachter, donated the bell, purchased for $35 by plumbing inspector John Poole during the 1945 razing of Second Ward City Hall. (Courtesy AFD Archive.)

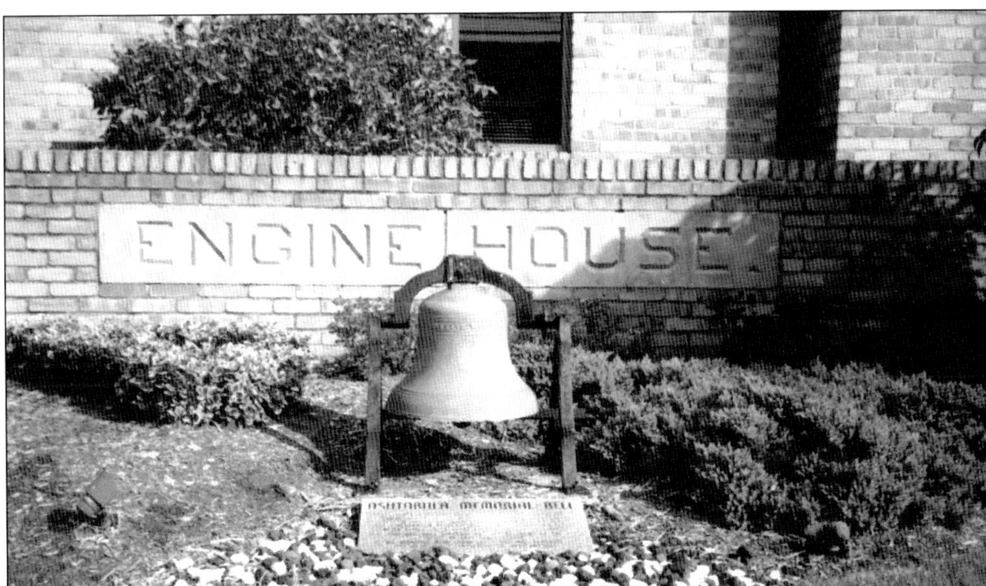


**Ashtabula Bridge Disaster Memorial, 2005.** In October 1999, the Ashtabula Fire Department dedicated this granite memorial marker near its c. 1854 Ashtabula Memorial Bell in a landscaped garden outside headquarters Station One. Donated and engraved by Ashtabula-based William Smith and Son Monument Company, the marker pays tribute to the 92 victims of the December 29, 1876, Ashtabula Bridge Disaster, still one of the nation's deadliest train disasters. (Author's collection.)

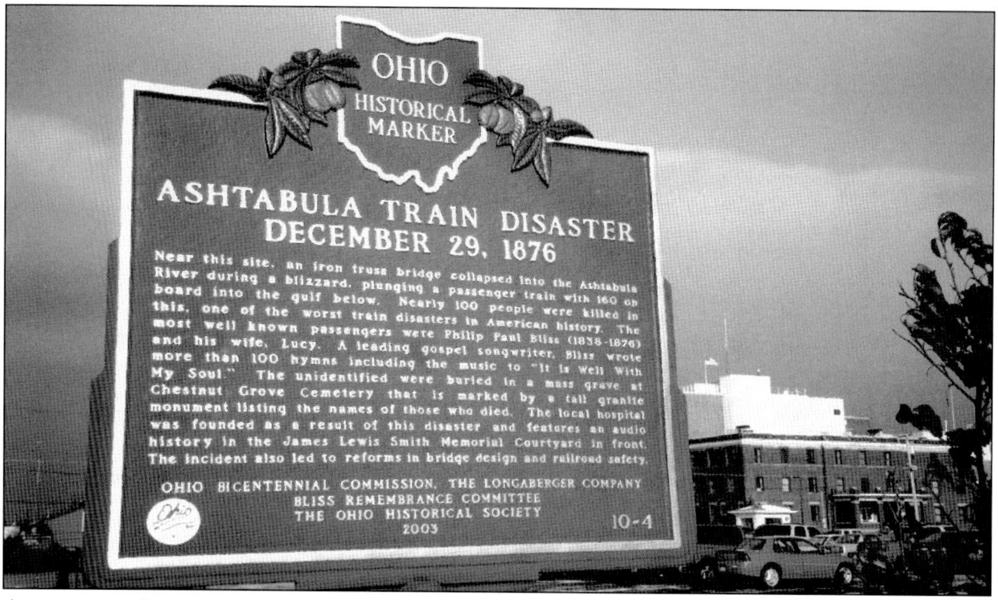


**ASHTABULA BRIDGE DISASTER HISTORIC MARKER, 2420 LAKE AVENUE.** Cooperative efforts of the Ohio Bicentennial Commission, Longaberger Company, Ashtabula's Bliss Remembrance Committee, and the Ohio Historical Society saw this Ohio Historical Marker dedicated in 2003, just north of the Ashtabula Bridge Disaster site. The December 1876 incident, which underscored the need for an Ashtabula-based hospital, spurred creation of a rudimentary emergency hospital clinic in 1882 and Ashtabula General Hospital in 1904. (Author's collection.)

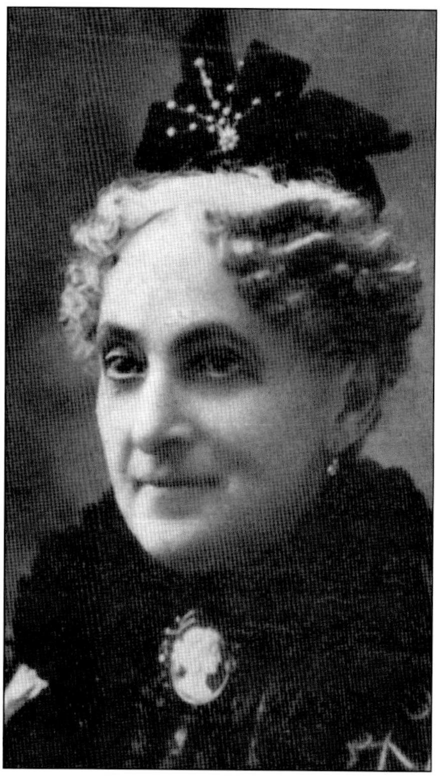

**HATTIE A. SHERMAN, 1902.** A native of Ashtabula's Bunker Hill neighborhood, Hattie Sherman was an active leader in city charitable work, establishing a Seamen's Bethel at Ashtabula Harbor and serving as an "indefatigable worker for the temperance cause." Sherman, through her "insistent demands and personal labor," was feted as a leading force behind the 1904 founding and construction of Ashtabula General Hospital, today's Ashtabula County Medical Center (ACMC). (Courtesy AFD Archive.)

# *Three*

# "PROFESSIONAL SERVICE SINCE 1911"

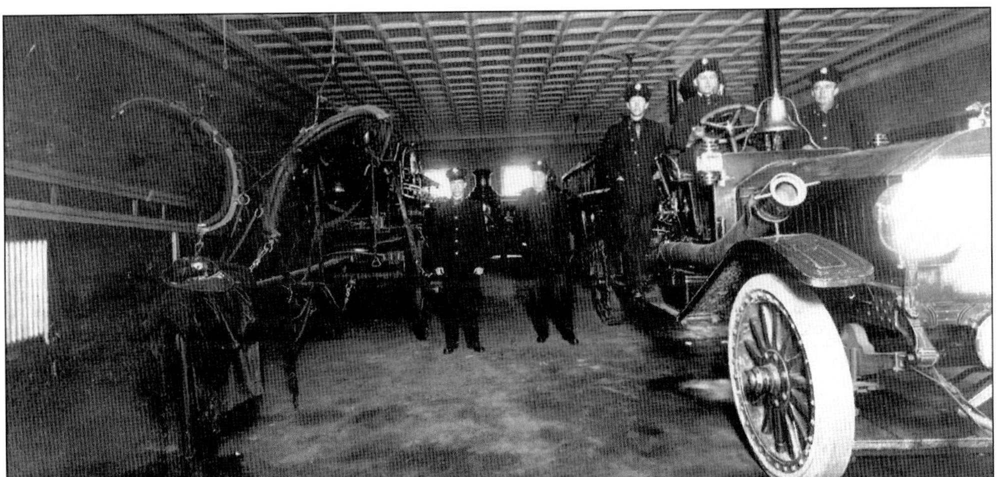

"ASHTABULA'S FIRST PAID FIRE COMPANY." As the city grew, and with it its firefighting needs, the Ashtabula Fire Department switched its operations from a volunteer force to a paid professional fire department. Starting with the Ashtabula Harbor fire station at 1123 Bridge Street in October 1911, the department's paid professional operations expanded in February 1912 to include the headquarters Uptown Fire Station at 4758 Park Avenue (pictured). Although a volunteer force prior to 1911, some department officers had been paid a regular annual salary as early as 1874. Concurrent with the historic operational changes, the department added its first pieces of motorized fire apparatus, including this state-of-the-art c. 1911 Robinson Jumbo pumper. Offering 110 horsepower, the piston-pump Robinson could pump 750 gallons per minute and shoot water up to 250 feet. And as a chemical engine, the Robinson used a soda-acid chemical reaction in its 30-gallon chemical tank to boost hose pressure. A time of transition, horse-drawn firefighting apparatus (left) remained an important part of firefighting operations, as illustrated by the harnesses attached to the department's horse-drawn aerial ladder truck. (Courtesy AFD Archive.)

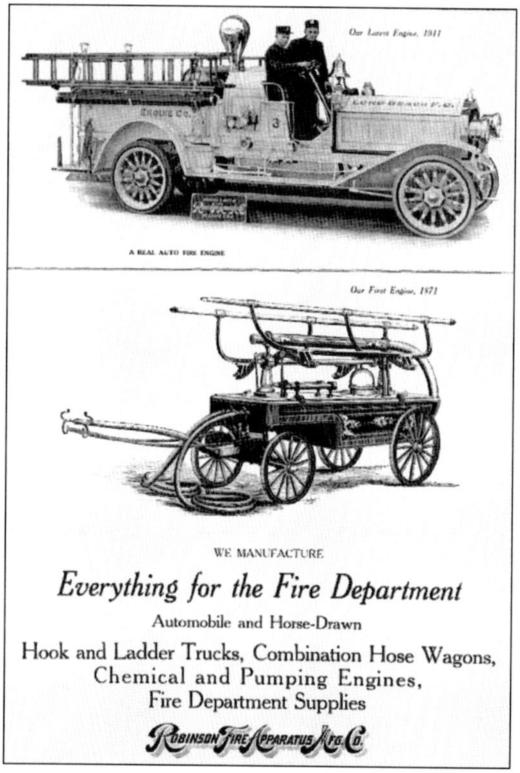


**ROBINSON ADVERTISEMENT, 1911.** Ashtabula joined numerous Ohio fire departments in purchasing a trailblazing auto fire engine from St. Louis–based Robinson Fire Apparatus Manufacturing Company (1871–1923), a premier early auto fire truck manufacturer. Billed as "wonderfully efficient, dependable, and economical," Robinson Jumbos were used in Ashtabula, Canton, Cleveland, Alliance, East Liverpool, and Youngstown, in addition to major United States cities including Detroit, St. Louis, Boston, Chicago, and Los Angeles. (Author's collection.)

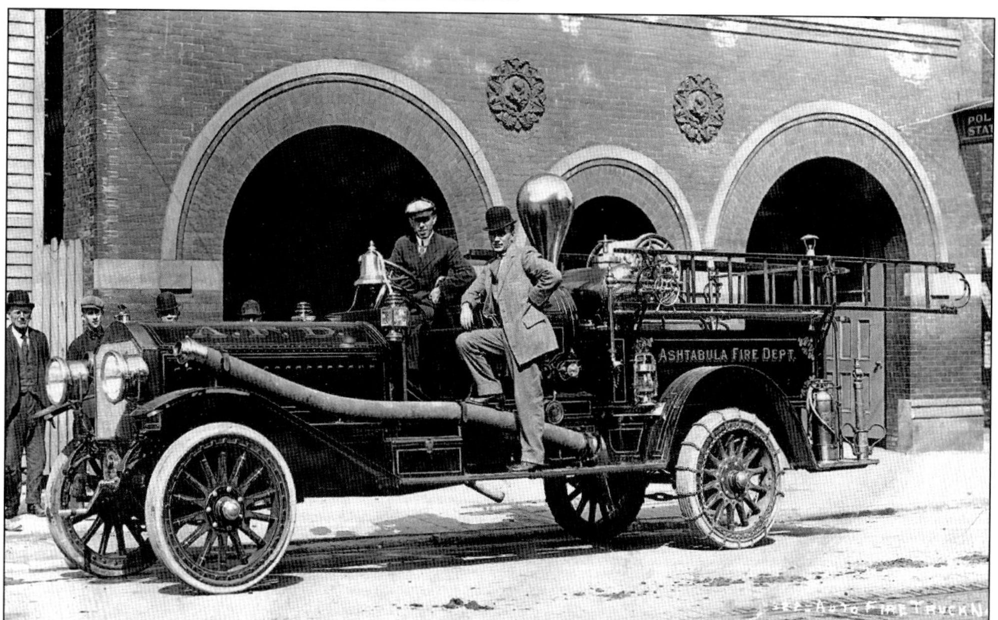


**ROBINSON JUMBO PUMPER, 1911.** This real-photo postcard shows firefighter Bill Hagrice (left) and then-driver Charles L. Devereaux on the Ashtabula Fire Department's newly delivered Robinson Jumbo auto fire truck outside the Harbor Fire Station, 1123 Bridge Street. Wrote Devereaux, "This is our new fire engine auto. I am one of the drivers at present. Don't know how it will be after the election. It is a dandy machine." (Author's collection.)

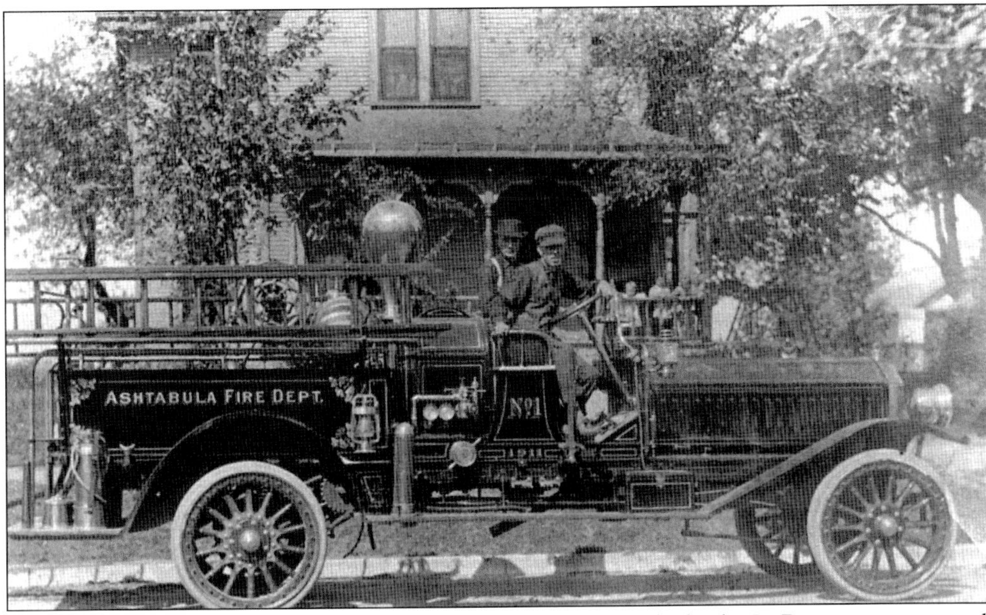


**POSING WITH THE NEW PUMPER.** Firefighters Bill Hagrice and Charles L. Devereaux pose with the department's new Robinson Jumbo pumper outside 163 Walnut Boulevard in Ashtabula Harbor. Looking on from the porch are children Samuel C., Edward L., and Edith B. "Pea" Pennoyer. Until the February 1912 opening of downtown's Uptown Fire Station, the Robinson was stationed at the Harbor Fire Station on Bridge Street. (Courtesy AFD Archive.)

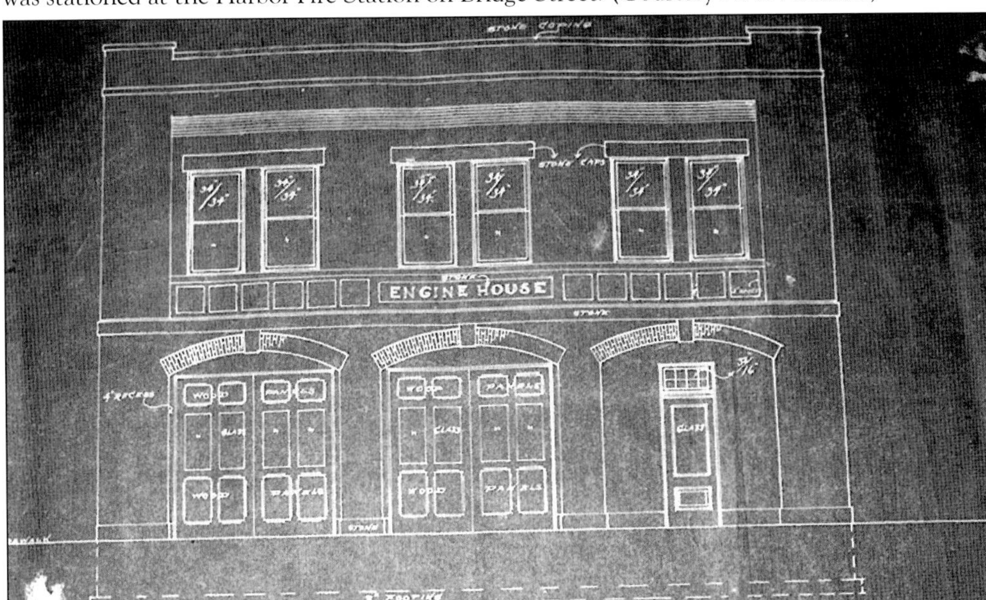


**UPTOWN BLUEPRINT, 1911.** This fragile leather-wrapped blueprint by Ashtabula architect William M. Johnson shows the front elevation of the Ashtabula Fire Department's headquarters Uptown Fire Station (1912–1975). Built downtown at Park Avenue and West Forty-eighth (Fisk) Street, Uptown's two-story Park Avenue facade featured a simple brick and stone face, highlighted by stylish wood panel and glass bay doors. The $7,499 station, the *Beacon-Record* noted, was "modern in every detail." (Courtesy AFD Archive.)

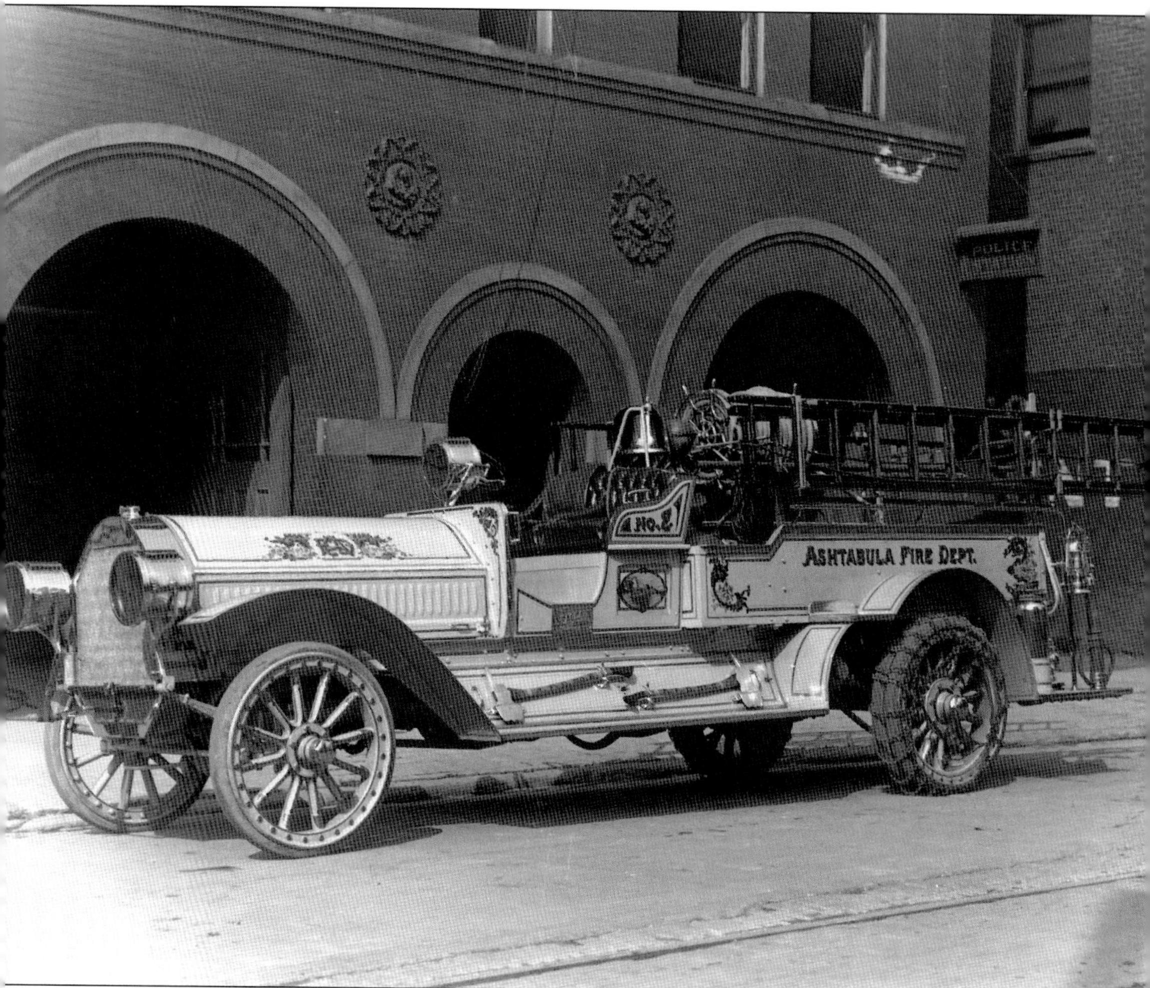


**SEAGRAVE HOSE AND CHEMICAL TRUCK, 1911.** Ashtabula's new $3,900 Seagrave Corporation hose and chemical engine, painted in light gray, sits prominently outside the Harbor Fire Station at 1123 Bridge Street. The driver's seat sported two decorative outer panel paintings of local scenes––an ore freighter plying Lake Erie (shown) and a Harbor scene of the tug *William D.* and the steamer *Alex Thompson*. Note the tire chains mounted on the solid rubber rear drive wheels, a year-round concession to the often-crude state of city roadways. Built at Seagrave's High Street manufacturing complex in Columbus, construction of Ashtabula's new hose and chemical engine, serial No. 5858, would "set the trend that would continue for years" as Seagrave's first water-cooled gable hood truck. Dubbed Engine Two, the six-cylinder hose and chemical truck, sporting an aluminum-coated steel body, was built with the best available components from leading suppliers including Bosch, Rayfield, Hancock, Timken, and Firestone. Said the Ashtabula *Beacon-Record* of the Seagrave, "The big fellow . . . is one of the finest ever put out from the state capital." (Courtesy Ashtabula Marine Museum.)

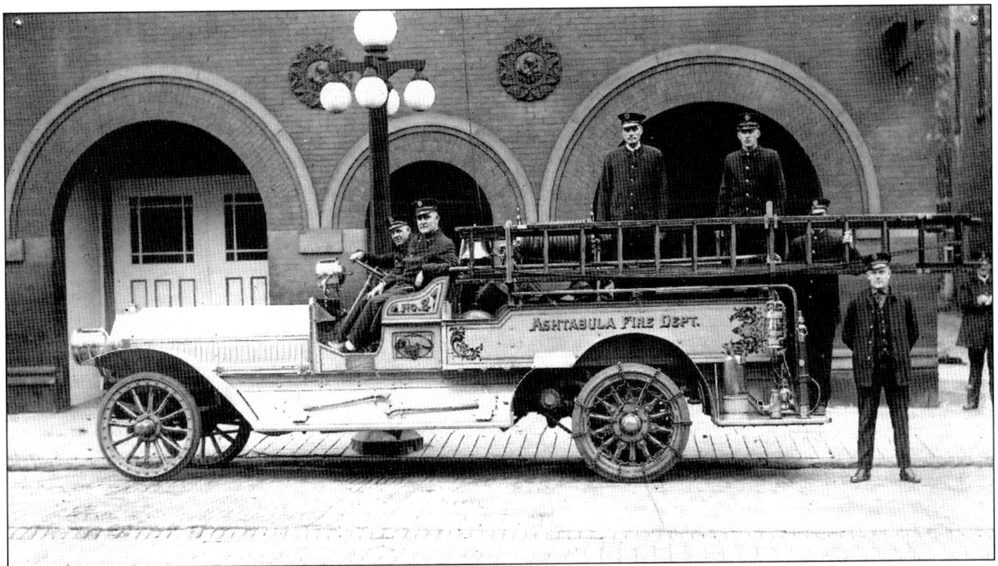


**SEAGRAVE HOSE AND CHEMICAL TRUCK, 1912.** Early Harbor Fire Station firefighters (from left to right) Carl Amsden, George H. Henniger, Steve Morris, J. F. Mullen, Dick Nelson (behind ladder), and future fire chief James M. Doran pose with Ashtabula's *c.* 1911 Seagrave hose and chemical truck outside the Bridge Street station. Purchase of the Seagrave, along with a 1911 Robinson Jumbo pumper, hastened the end of the department's horse-drawn era. (Courtesy AFD Archive.)

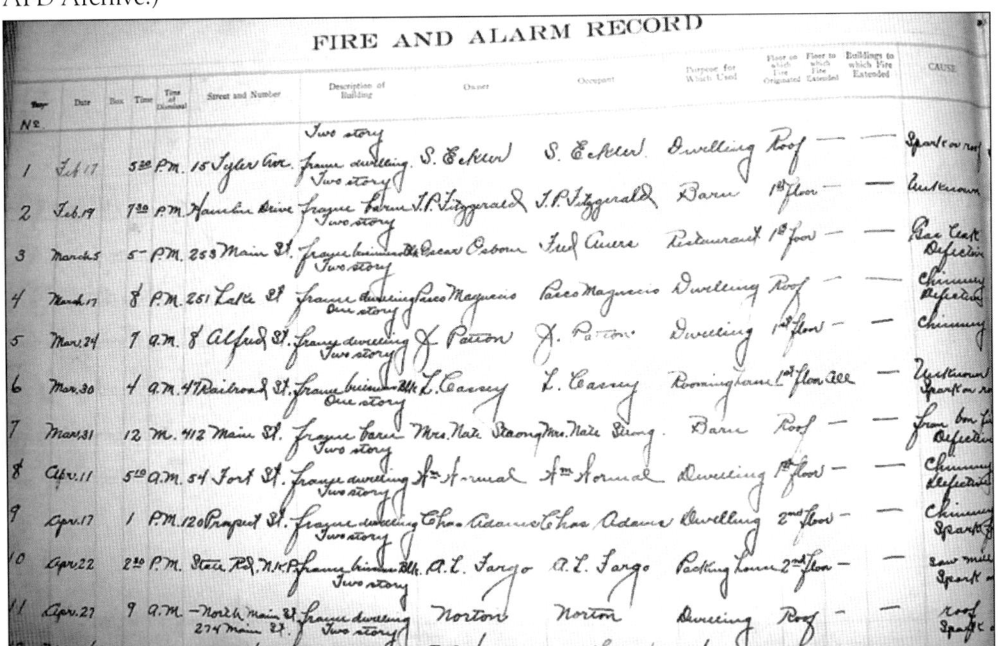


**UPTOWN FIRE LOG, 1912.** Handwritten log entries chronicle initial runs made from Ashtabula's new headquarters Uptown Fire Station, located at 4758 Park Avenue. Opened February 12, 1912, under mayor I. H. Pardee, safety director B. S. Bliss, and fire chief George E. Ducro Sr., charter Uptown staff were Capt. Charles L. Devereaux, Lt. Fred D. Alexander, driver Archie Castle, and firemen Dennis T. Leahy, James Dunlavy, and John D. Huggins. (Courtesy AFD Archive.)

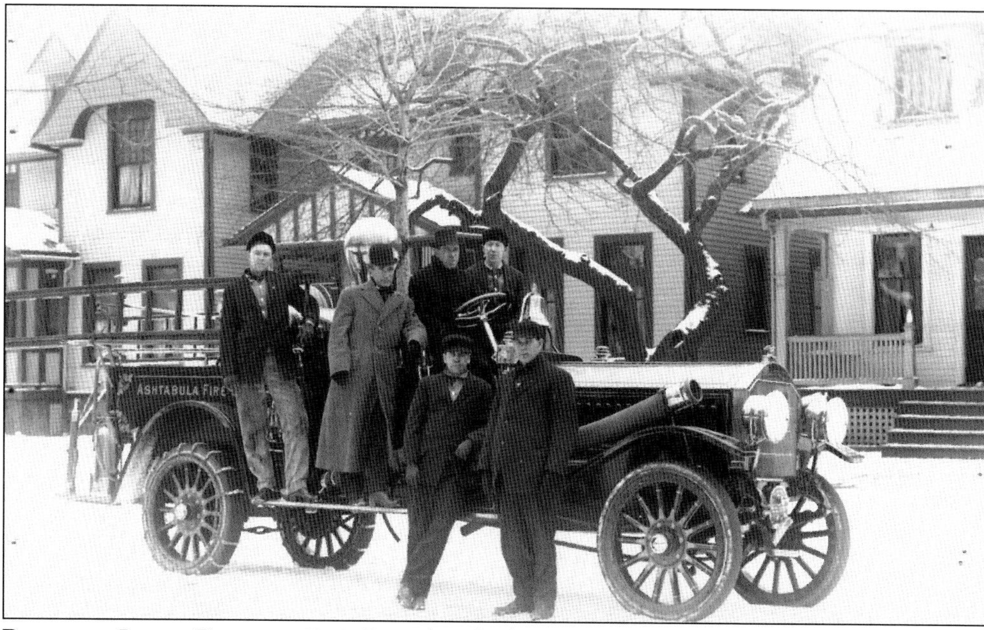


**ROBINSON JUMBO PUMPER, 1912.** Posted to the headquarters Uptown Fire Station, this *c.* 1911 Robinson Jumbo pumper was one of Ashtabula's first auto fire trucks, purchased in conjunction with the formation of the city's paid professional fire department. Pictured here from left to right are (on the Robinson) unidentified, Charles L. Devereaux, William (Bill) Haigney, and Fred D. Alexander; (on the ground) Charles E. Lasky and Daniel Condron. (Courtesy AFD Archive.)

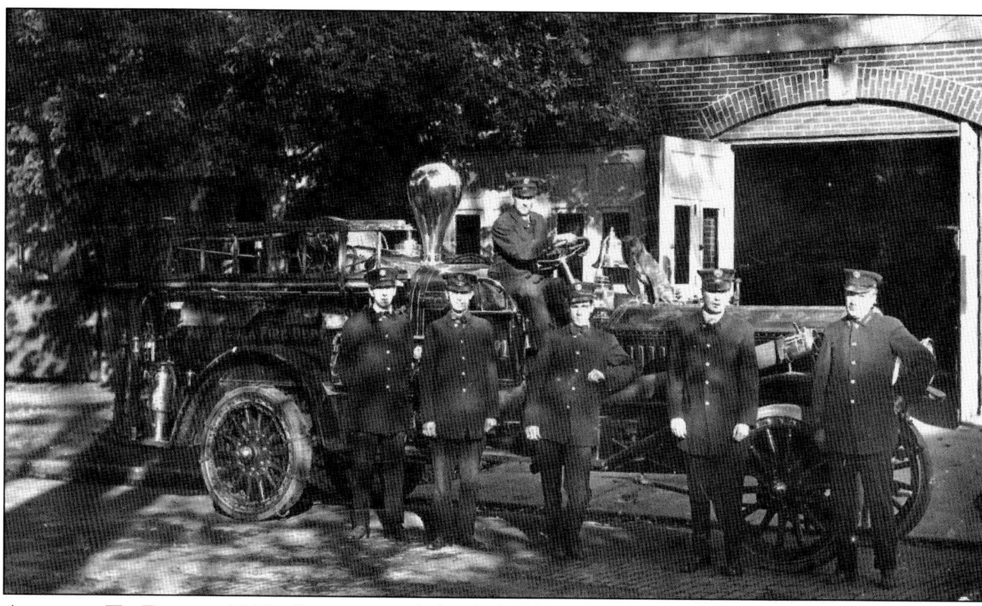

**ADOLPH T. ROHL, 1913.** Longtime Ashtabula fire chief and 50-year department veteran Adolph T. Rohl (center front) is seen as a rookie 21-year-old Uptown hoseman in this real-photo postcard. Pictured standing to the left of Rohl are Charles E. Lasky and Daniel Condron. Standing to Rohl's right are future city fire chiefs Capt. Charles L. Devereaux and Lt. Fred D. Alexander. William Haigney is the driver. (Courtesy AFD Archive.)

**HARRY C. FROID, 1913.** Following in the footsteps of his father, an immigrant Swedish fireman, Ashtabula-born firefighter Harry C. Froid is just beginning his longtime career with the Ashtabula Fire Department. Joining Ashtabula's fledgling professional fire department at age 21 in April 1913, Froid worked his way up through Uptown's ranks, appointed driver in 1919 and lieutenant in 1946. Health concerns spurred 48-year veteran Froid's June 1961 retirement. (Courtesy AFD Archive.)

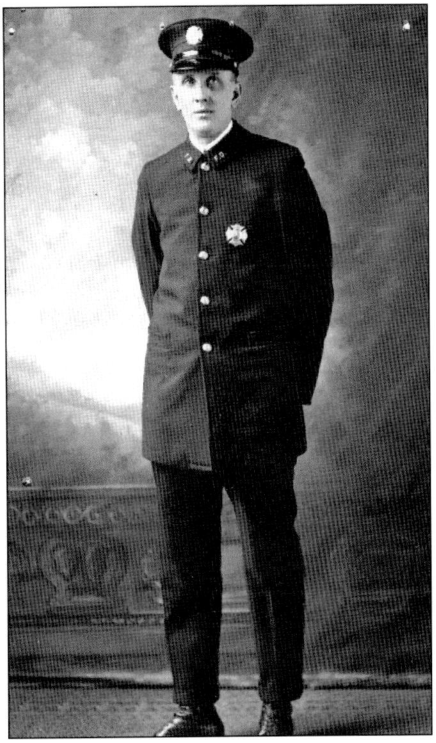

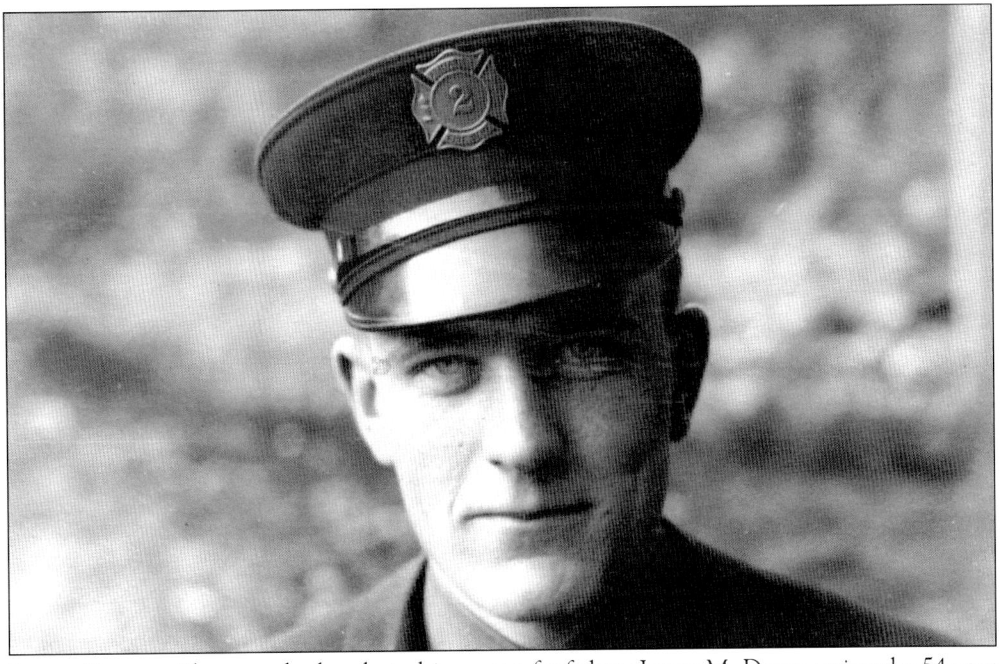

**JIMMY DORAN.** Photographed early in his career, firefighter James M. Doran enjoyed a 54-year tenure with the Ashtabula Fire Department, joining as a hoseman in July 1917 and retiring as fire chief in May 1971. A key figure in the 1919 drive to organize Ashtabula's International Association of Fire Fighters (IAFF) Local 165, Doran would serve Local 165 as president and vice president, logging 15 years as secretary. (Courtesy AFD Archive.)

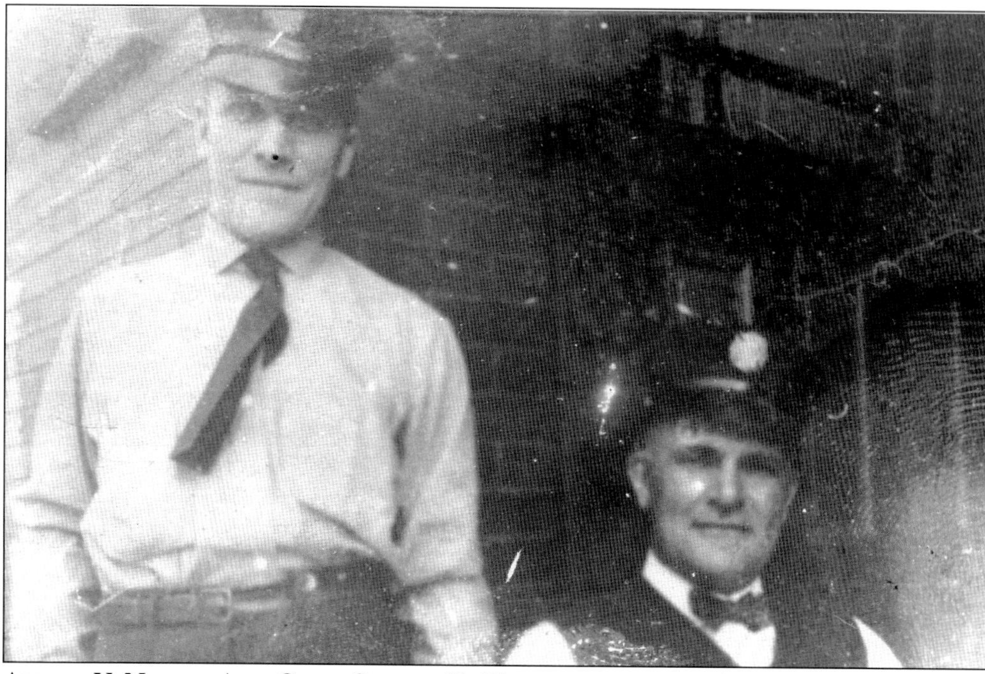

**ALBERT V. NELSON AND CAPT. GEORGE H. HENNIGER, 1914.** Harbor Fire Station firefighter Albert V. Nelson (left) is just starting his 43-year career, which ran from May 1914 to January 1957. At right is department veteran Capt. George H. Henniger. Other Harbor firefighters included engineers Carl C. Amsden and Charles E. Johnson, Lt. Charles E. Lasky, and hosemen Steven A. Morris, Otto L. Collar, and Charles J. Lloyd. (Courtesy AFD Archive.)

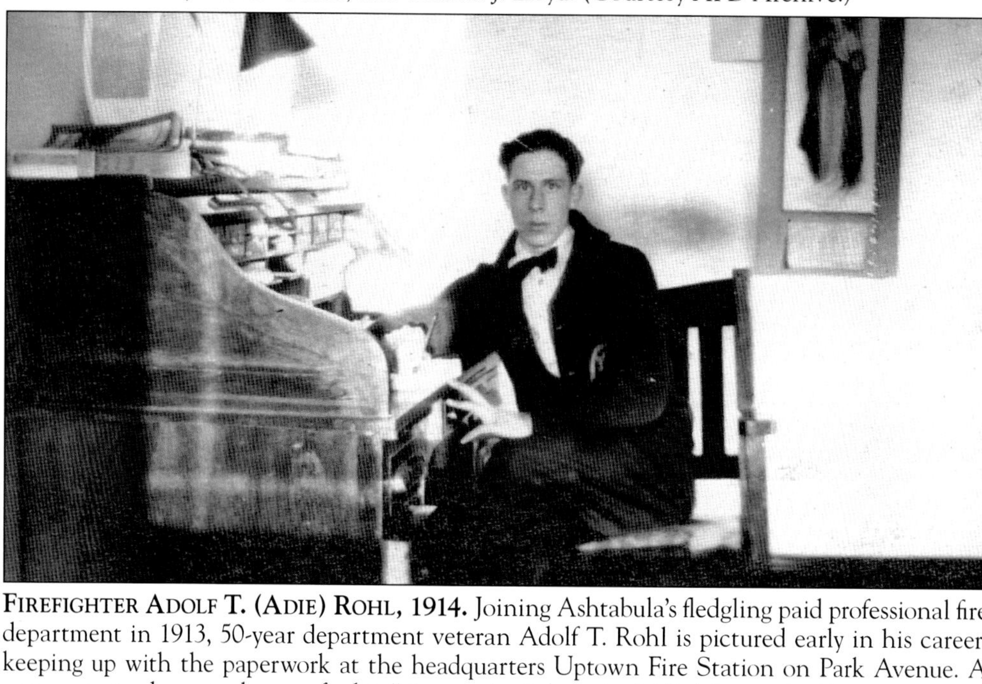

**FIREFIGHTER ADOLF T. (ADIE) ROHL, 1914.** Joining Ashtabula's fledgling paid professional fire department in 1913, 50-year department veteran Adolf T. Rohl is pictured early in his career, keeping up with the paperwork at the headquarters Uptown Fire Station on Park Avenue. A consummate character known for his "cigar antics," the colorful Rohl served a 19-year tenure as fire chief from 1944 to 1963. (Courtesy AFD Archive.)

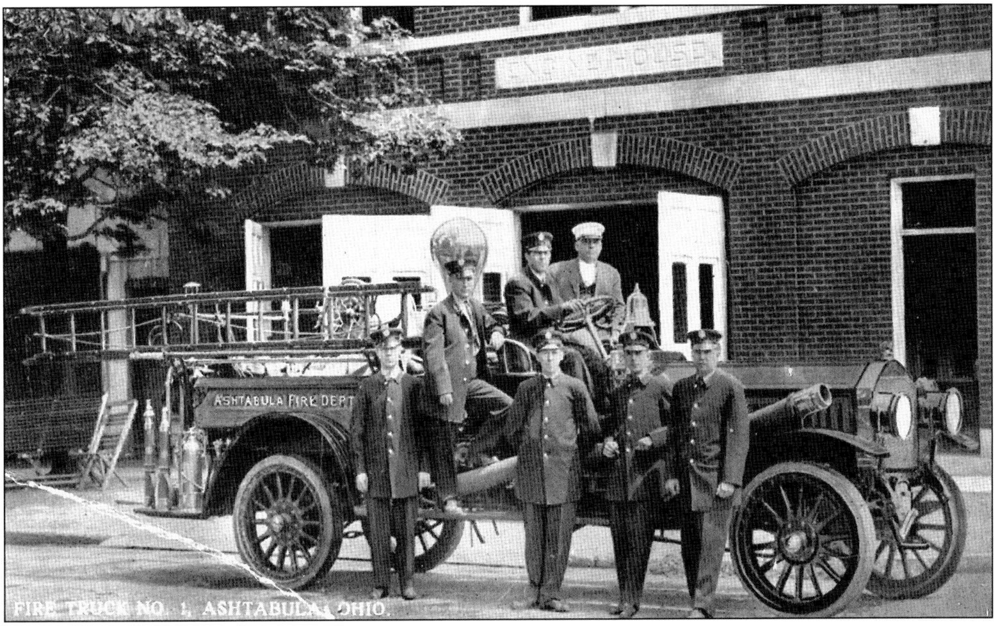


**ENGINE 1, 1915.** With the 1912 debut of the headquarters Uptown Fire Station, Ashtabula was divided at Nineteenth Street into northern and southern quadrants served by the Harbor and Uptown Stations respectively. Shown here, from left to right, are (riding on Engine No. 1) Capt. Charles L. Devereaux, driver Archie Castle, and fire chief George E. Ducro Sr.; (standing) John D. Huggins, Lt. Fred D. Alexander, James Dunlavy, and Charles E. Lasky. (Courtesy AFD Archive.)

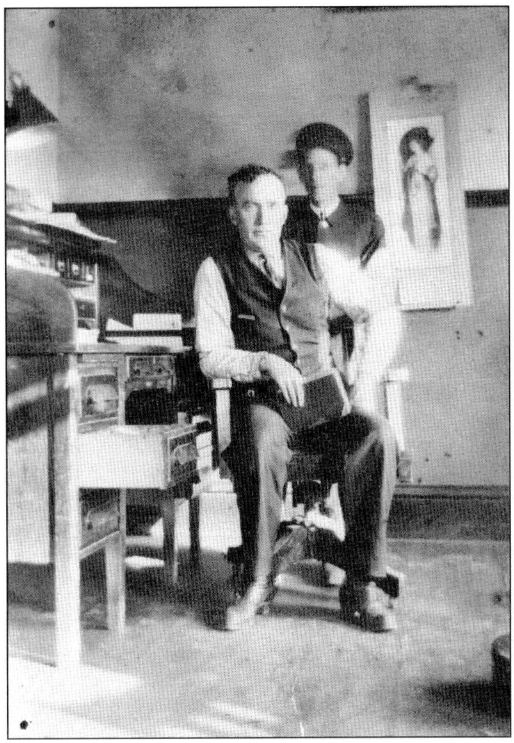

**GEORGE E. DUCRO SR., MID-1910S.** Shown are future 1944–1963 fire chief Adolph T. Rohl (back) and then–fire chief George E. Ducro Sr. (1907–1916). A pivotal chief, Ducro spearheaded creation of the paid professional fire department, construction of the Uptown Fire Station, and the transformation to a motorized firefighting force. Admired Rohl, "The best fireman I ever served under was George Ducro. He was . . . the best 'smoke eater' I've ever seen." (Courtesy AFD Archives.)

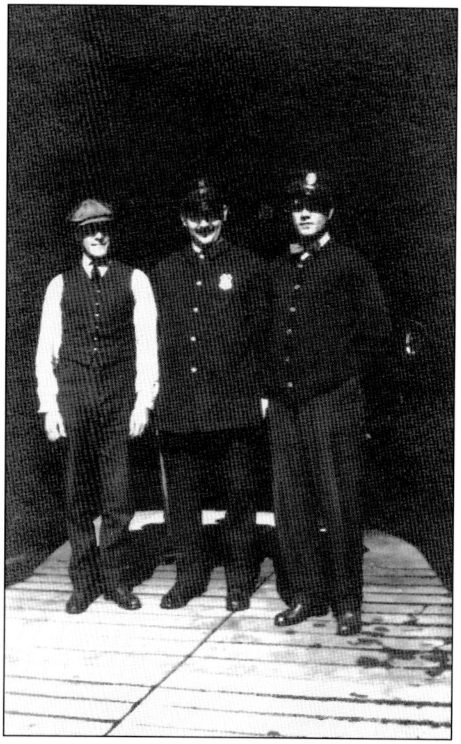

**PUBLIC SERVANTS.** Ashtabula Police Department patrolman Samuel A. Clint (center) visits firefighters Albert V. Nelson (left) and Arthur C. Larson at the Harbor Fire Station on Bridge Street. Seeking to improve emergency response times, Ashtabula converted many of its locally significant east-west street names to numbered street names in 1930. The old historic street names still unofficially endure among longtime residents. (Courtesy AFD Archive.)

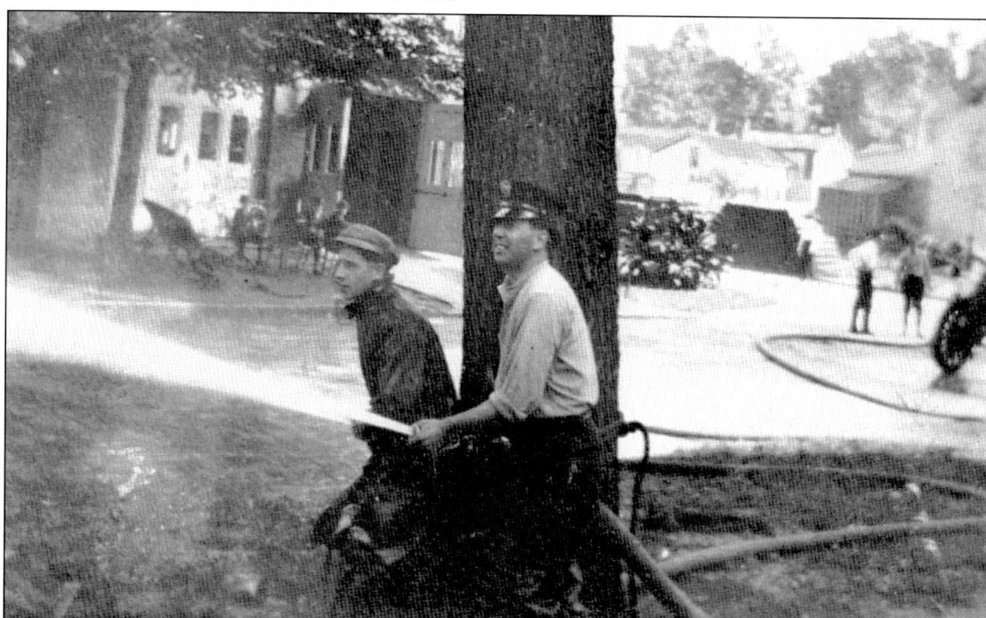

**HOSE DRILLS.** City firefighters conduct hose drills with the *c.* 1911 Robinson Jumbo pumper outside the headquarters Uptown Fire Station on Park Avenue in downtown Ashtabula. The chain-drive pumper, a technological marvel of its day, was built by St. Louis–based Robinson Fire Apparatus Manufacturing Company, one of the nation's premier early motorized fire apparatus manufacturers between 1908–1923. Decoratively, Robinsons were distinguished by their perching eagle radiator caps. (Courtesy AFD Archive.)

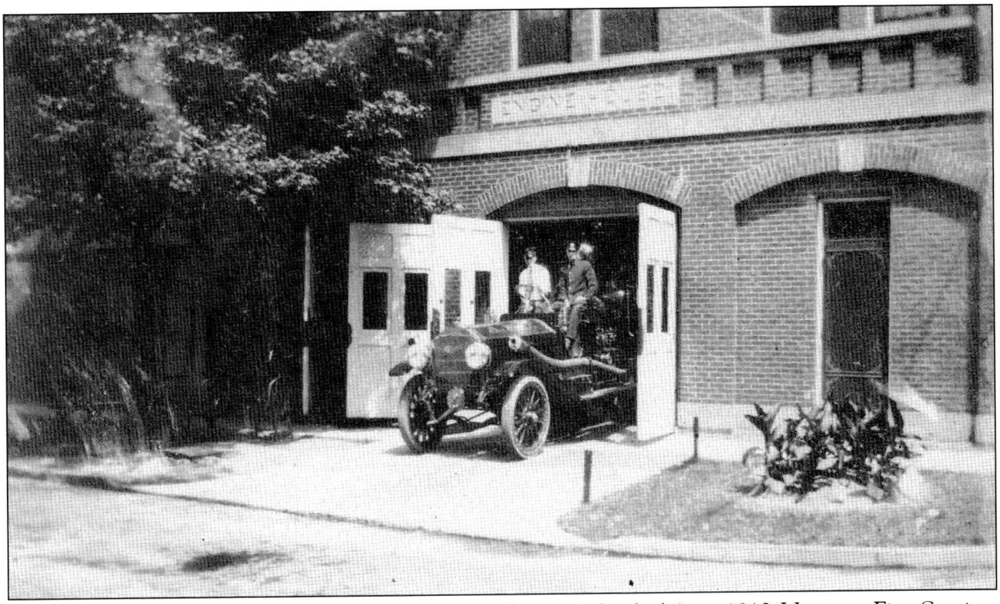

**"ASHTABULA FIRE STATION."** This rare view shows Ashtabula's *c.* 1912 Uptown Fire Station with its original decorative Park Avenue tree lawn. A later widening of Park Avenue claimed the tree lawn, leaving the station surrounded by concrete. At center are Uptown firefighters aboard their 1911 Robinson Jumbo pumper. Only a handful of 1911 Robinson Jumbos survive, including fire museum restorations in Gila Bend, Arizona, and Staunton, Virginia. (Courtesy AFD Archive.)

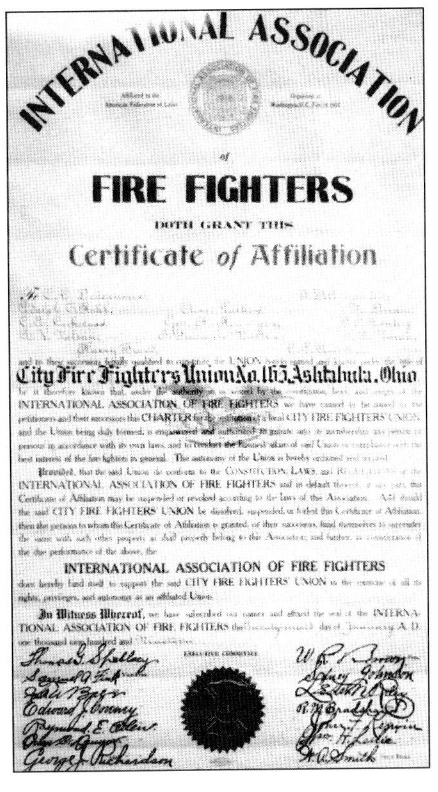


**IAFF LOCAL 165 CHARTER, JANUARY 22, 1919.** Ashtabula firefighters were among the nation's first to join the *c.* 1918 IAFF. Organized by the American Federation of Labor, IAFF tackled firefighter issues including long workweeks, low salaries, non-existent pensions, frequent firefighter fatalities, and a lack of adequate safety equipment. Local 165 organizers included future fire chiefs Charles Devereaux, Fred Alexander, Adolph Rohl, and James Doran. (Courtesy AFD Archive.)

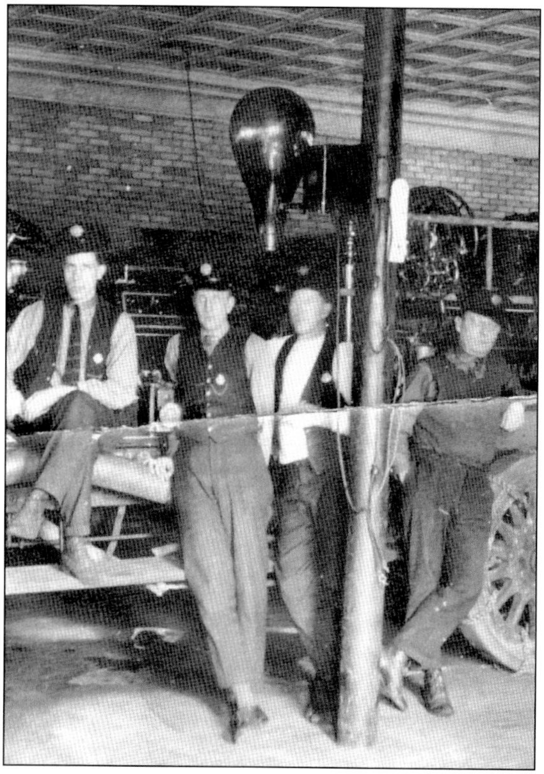

**UPTOWN FIRE STATION.** Early professional Ashtabula firefighters, from left to right, Adolph T. Rohl, Harry Froid, Lt. Fred D. Alexander, and Dennis T. Leahy pose inside Uptown's truck bay with their c. 1911 Robinson Jumbo pumper. Note the ornate tin ceiling, a common decorative treatment between the mid-1800s and World War I. A uniquely North American design element, economical tinplate panels were stamped with copies of fine carved and molded European plasterwork. (Courtesy AFD Archive.)

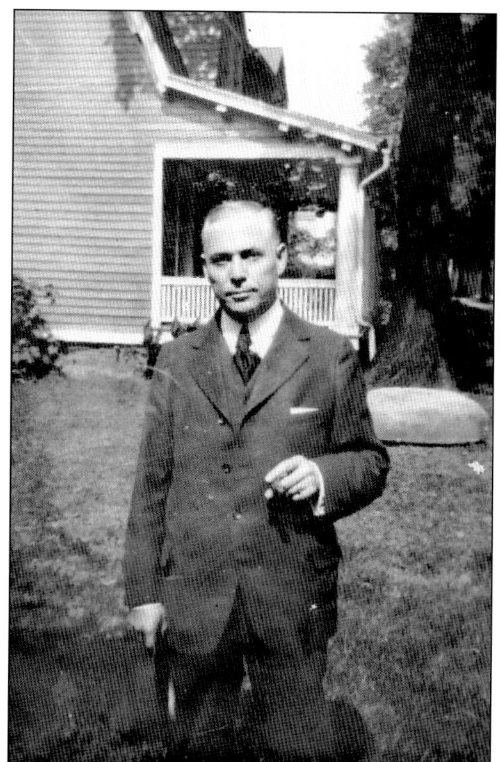

**GEORGE EDWARD DUCRO SR., 1919.** Prominent downtown furniture retailer George E. Ducro Sr. (1867–1957) was long active in city firefighting, first as a Protection Fire volunteer firefighter and officer, and later as city fire chief from 1907 to 1916. Long associated with the family business, Ducro Furniture Company (1853–1973), Rotarian Ducro also served on numerous director boards, including Farmers National Bank and Trust and the Hotel Ashtabula. (Courtesy George E. Ducro III.)

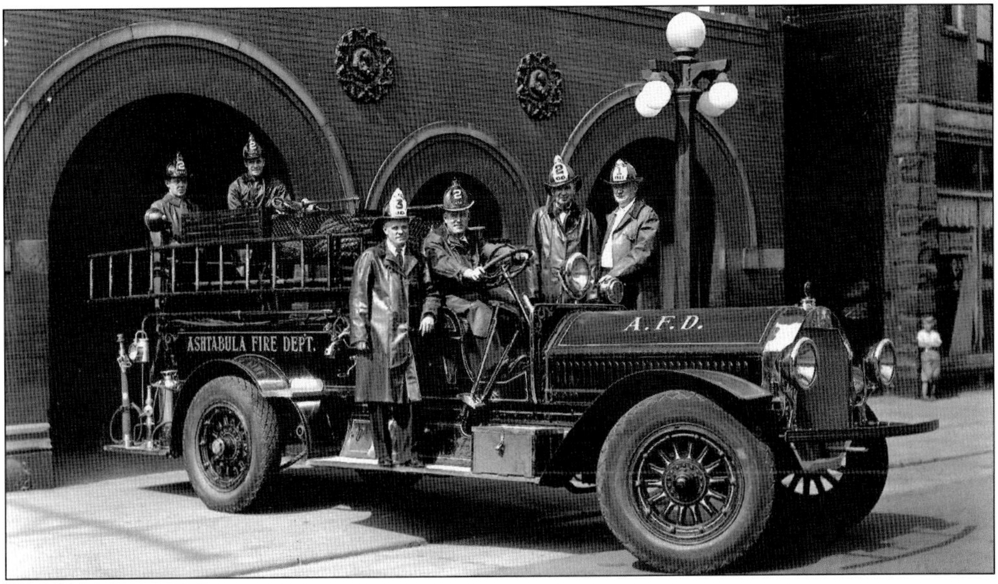


**NEW AMERICAN-LAFRANCE PUMPER.** The ebullient boom years of the Roaring Twenties saw the Ashtabula Fire Department purchase this 1922 American-LaFrance "Triple Combination" chemical, hose, and pumper engine, pictured outside the Harbor Fire Station on Bridge Street. Pictured here, from left to right, are hosemen Charles Lloyd and Albert V. Nelson, Lt. Charles Johnson, driver Arthur C. Larson, Lt. Charles E. Lasky, and Capt. Charles L. Devereaux, city fire chief from 1939 to 1943. (Courtesy AFD Archive.)

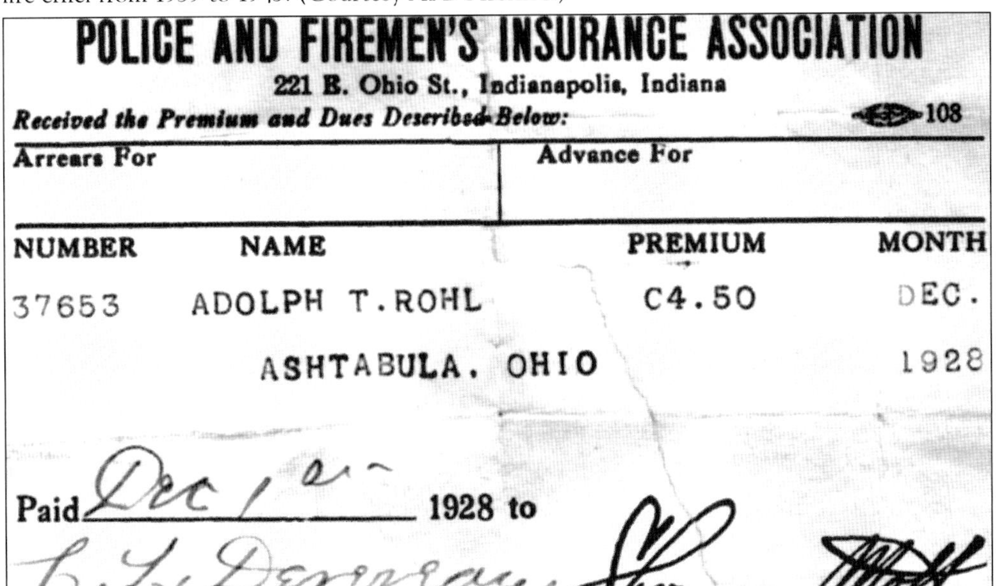

POLICE AND FIREMEN'S INSURANCE ASSOCIATION

221 B. Ohio St., Indianapolis, Indiana

*Received the Premium and Dues Described Below:* 108

| Arrears For | Advance For |
|---|---|
| | |

| NUMBER | NAME | PREMIUM | MONTH |
|---|---|---|---|
| 37653 | ADOLPH T.ROHL | C4.50 | DEC. |
| | ASHTABULA, OHIO | | 1928 |

Paid Dec 1st 1928 to

C. L. Devereaux Sherman Mott

**POLICE AND FIREMEN'S INSURANCE ASSOCIATION RECEIPT, 1928.** Among the benefits of IAFF membership was eligibility for various insurance benefits. Pictured is engineer Adolph Rohl's December 1928 receipt for his monthly insurance premium, signed by Ashtabula Local 165 advisory board member Capt. Charles L. Devereaux. Serving the department between 1913 and 1963, Rohl was promoted to lieutenant in 1930, captain in 1939, and fire chief in 1945. (Courtesy AFD Archive.)

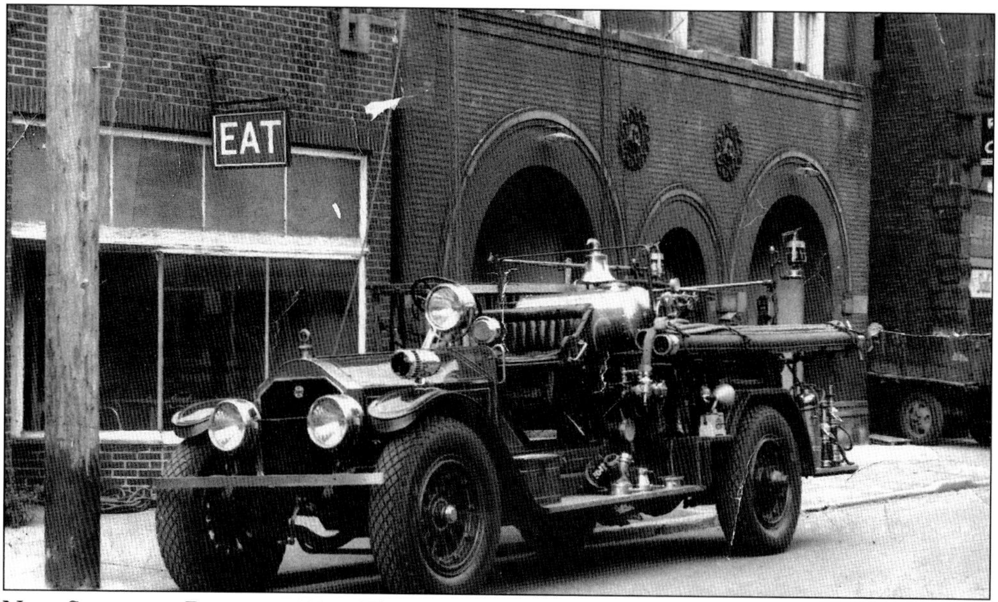


**New Seagrave Pumper, 1928.** The economic prosperity of the Roaring Twenties saw the Ashtabula Fire Department add several new units, including this c. 1928 Seagrave combination booster tank, hose, and pumper engine, pictured outside the Harbor Fire Station on Bridge Street. In the early months of the Great Depression, the department added a much-needed 75-foot c. 1930 Seagrave aerial hook and ladder truck, its final major apparatus acquisition until 1944. (Courtesy AFD Archive.)

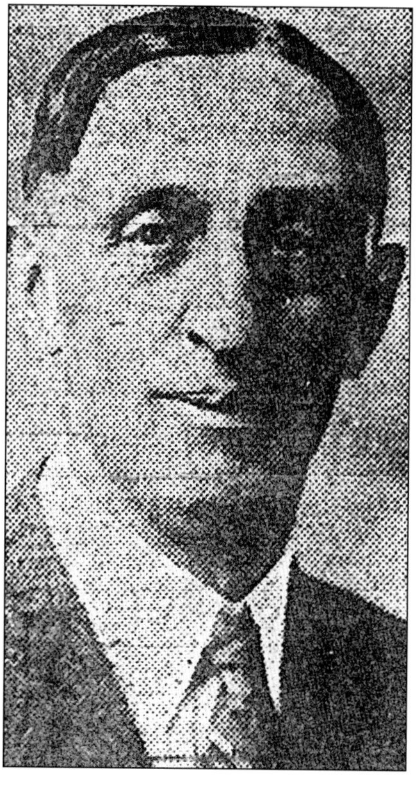

**Ray N. Mahan, 1929.** After 13 years without a fire chief, city manager Fred R. Hogue appointed 1897–1912 Protection Fire volunteer and former assistant fire chief Ray Mahan as Ashtabula's first full-time professional fire chief in 1929. Known during his eight-year tenure for his "war on hazards" emphasis on fire prevention, Mahan resigned in 1937 to become an inspector with Middle Western Fire Inspection Company in the Mahoning Valley. (Courtesy Ashtabula County District Library.)

**UPTOWN LOG BOOK, APRIL 1930.** Uptown Fire Station calls were recorded by hand in 1930, with notations made on the run, responding department officers, and apparatus used. Within a two-day period, busy Uptown firefighters had made two runs to the Palace Theater in the 4700 block of Main Avenue, then 242 Main Avenue, once for a bomb blast and once for a fire ignited by an improperly discarded cigarette. (Courtesy AFD Archive.)

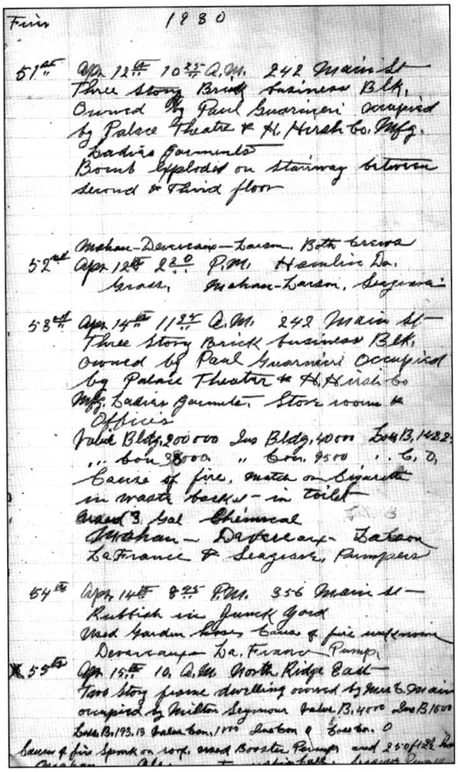

Fires 1930

51st Apr. 12th 10.25 A.M. 242 Main St. Three story Brick business Blk. Owned by Paul Guarinieri occupied by Palace Theatre & H. Hirsh Co. Mfg. Ladies Garments. Bomb Exploded on stairway between second & third floor. Mahan–Devercaux–Larson. Both crews

52nd Apr. 12th 2.20 P.M. Hamlin Dr. Grass. Mahan–Larson, Seagrave

53rd Apr. 14th 11.25 A.M. 242 Main St. Three story Brick business Blk. owned by Paul Guarinieri Occupied by Palace Theatre & H. Hirsh Co. Mfg. Ladies Garments. Store rooms & Offices. Value Bldg. 200 000 Ins Bldg. 40 000 Loss B. 1422 ; Con. 38000 Con. 9500 C. 0. Cause of fire, match or cigarette in waste basket – in toilet. Used 3 Gal Chemical. Mahan – Devercaux – Larson. LaFrance & Seagrave, Pumpers

54th Apr. 14th 8.25 P.M. 356 Main St. Rubbish in Junk Yard. Used Garden Hose. Cause of fire unknown. Devercaux – La France Pump.

55th Apr. 15th 10. A.M. North Ridge East. Two story frame dwelling owned by Mrs. E. Main occupied by Milton Seymour. Value B. 4000 Ins B. 1500 Loss B. 193.13 Value Con. 1000 Ins Con. 0 Loss Con. 0. Cause of fire Spark on roof. Used Booster Pump and 250 ft 1½ hose

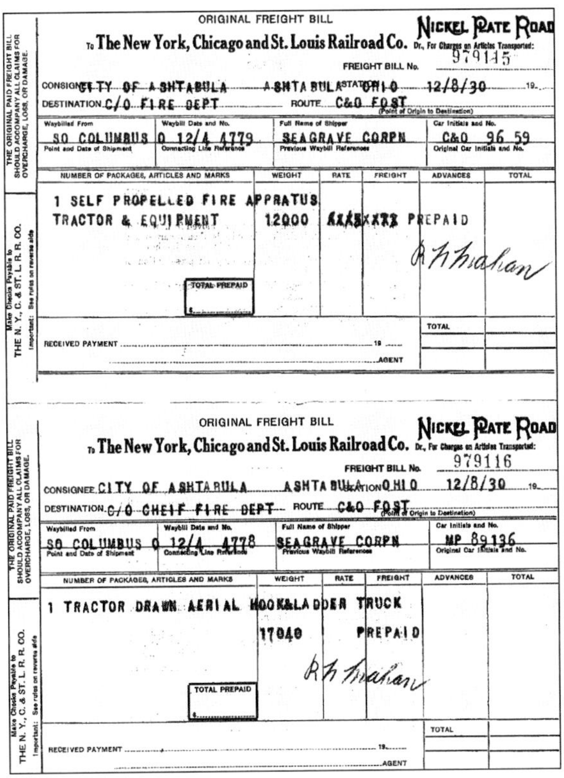

ORIGINAL FREIGHT BILL

NICKEL PLATE ROAD

To The New York, Chicago and St. Louis Railroad Co. Dr., For Charges on Articles Transported:

FREIGHT BILL No. 979115

CONSIGNEE CITY OF ASHTABULA ASHTABULA STATION OHIO 12/8/30 19

DESTINATION C/O FIRE DEPT ROUTE C&O FOST (Point of Origin to Destination)

Waybilled From: SO COLUMBUS (Point and Date of Shipment) | Waybill Date and No.: 0 12/4 4779 (Connecting Line Reference) | Full Name of Shipper: SEAGRAVE CORPN (Previous Waybill Reference) | Car Initials and No.: C&O 96 59 (Original Car Initials and No.)

| NUMBER OF PACKAGES, ARTICLES AND MARKS | WEIGHT | RATE | FREIGHT | ADVANCES | TOTAL |
|---|---|---|---|---|---|
| 1 SELF PROPELLED FIRE APPRATUS TRACTOR & EQUIPMENT | 12000 | XXXXXXX | PREPAID | | |

TOTAL PREPAID $

R H Mahan

TOTAL

RECEIVED PAYMENT 19 AGENT

THE ORIGINAL PAID FREIGHT BILL SHOULD ACCOMPANY ALL CLAIMS FOR OVERCHARGE, LOSS, OR DAMAGE.

Make Checks Payable to THE N. Y., C. & ST. L. R. R. CO. Important: See rules on reverse side

ORIGINAL FREIGHT BILL

NICKEL PLATE ROAD

To The New York, Chicago and St. Louis Railroad Co. Dr., For Charges on Articles Transported:

FREIGHT BILL No. 979116

CONSIGNEE CITY OF ASHTABULA ASHTABULA STATION OHIO 12/8/30 19

DESTINATION C/O CHEIF FIRE DEPT ROUTE C&O FOST (Point of Origin to Destination)

Waybilled From: SO COLUMBUS (Point and Date of Shipment) | Waybill Date and No.: O 12/4 4778 (Connecting Line Reference) | Full Name of Shipper: SEAGRAVE CORPN (Previous Waybill Reference) | Car Initials and No.: MP 89136 (Original Car Initials and No.)

| NUMBER OF PACKAGES, ARTICLES AND MARKS | WEIGHT | RATE | FREIGHT | ADVANCES | TOTAL |
|---|---|---|---|---|---|
| 1 TRACTOR DRAWN AERIAL HOOK&LADDER TRUCK | 17040 | | PREPAID | | |

TOTAL PREPAID $

R H Mahan

TOTAL

RECEIVED PAYMENT 19 AGENT

THE ORIGINAL PAID FREIGHT BILL SHOULD ACCOMPANY ALL CLAIMS FOR OVERCHARGE, LOSS, OR DAMAGE.

Make Checks Payable to THE N. Y., C. & ST. L. R. R. CO. Important: See rules on reverse side

**NICKEL PLATE ROAD FREIGHT BILL, 1930.** Trading in Moore Hook and Ladder's old horse-drawn aerial wagon, the Ashtabula Fire Department took delivery on a Columbus-built 1930 Seagrave aerial ladder truck. Shipped on the Nickel Plate railroad, the state-of-the-art $30,000 unit encompassed a self-propelled motorized tractor and a 75-foot tractor-drawn aerial hook and ladder truck. The Seagrave aerial (1930–1966) is today owned by the Fire Historical Society of Ashtabula. (Courtesy AFD Archive.)

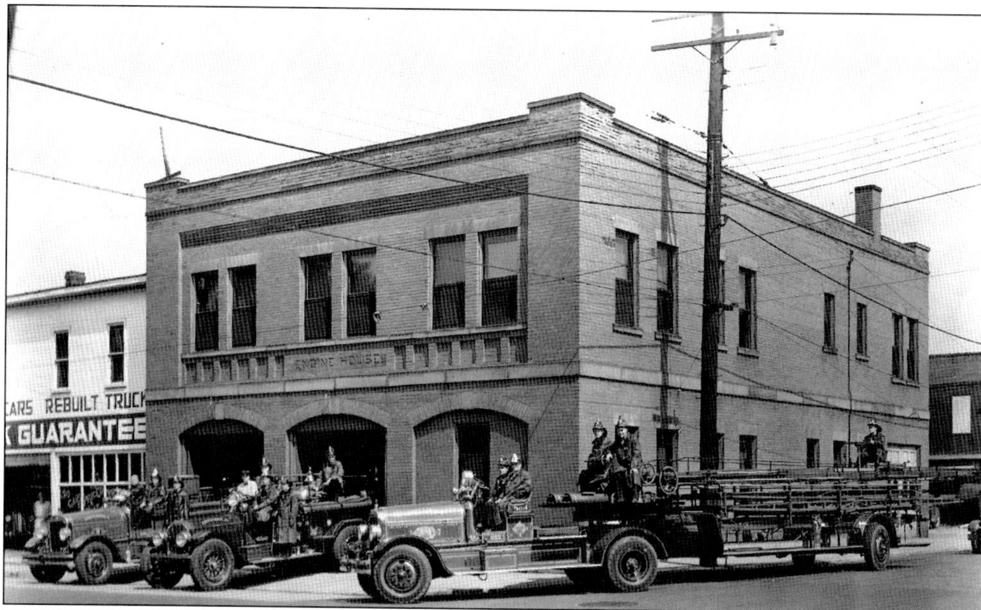


**Ashtabula Fire Department Anniversary, 1931.** Celebrating the paid professional fire department's 20th anniversary, Ashtabula Fire Department apparatus is staged outside the *c.* 1912 headquarters Uptown Fire Station at Park Avenue and West Forty-eighth (Fisk) Street in downtown Ashtabula. Pictured from left to right are the department's *c.* 1928 Seagrave pumper, its *c.* 1922 American-LaFrance pumper, and its 75-foot *c.* 1930 Seagrave aerial hook and ladder truck. (Courtesy AFD Archive.)

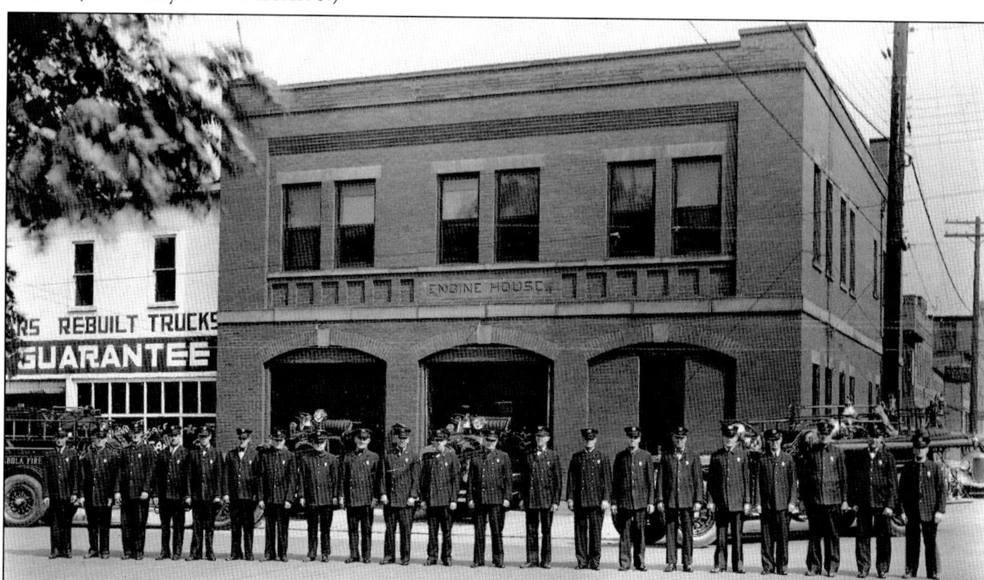


**Ashtabula Fire Department Anniversary, 1931.** Marking 20 years of paid professional firefighting, Ashtabula's 21 firefighters gather outside the Uptown Fire Station. Fire chief Ray N. Mahan's 1936 officer roster included Capts. Charles L. Devereaux, Fred D. Alexander, and G. H. Henniger, and Lts. Adolph T. Rohl, Arthur C. Larson, and Charles E. Lasky. Divided into three companies, firefighters worked a two-platoon system, 24 hours on duty and 24 hours off. (Courtesy AFD Archive.)

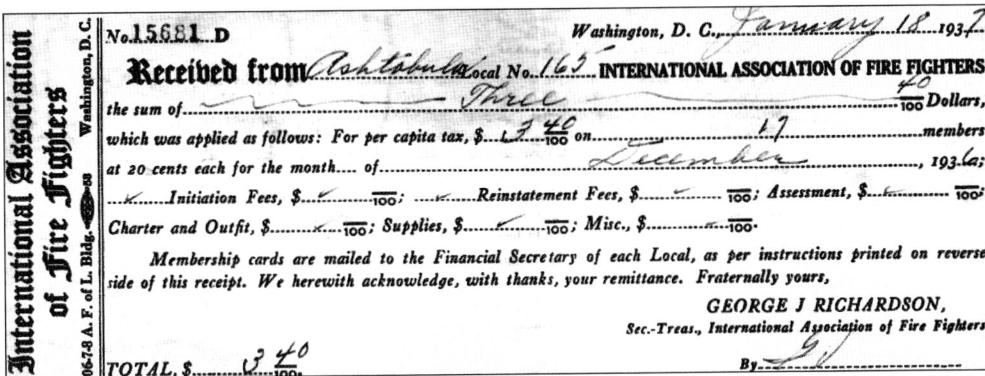

International Association of Fire Fighters
606-7-8 A. F. of L. Bldg. Washington, D. C.

No. 15681 D — Washington, D. C., January 18 1937

Received from Ashtabula Local No. 165 INTERNATIONAL ASSOCIATION OF FIRE FIGHTERS

the sum of Three 40/100 Dollars,

which was applied as follows: For per capita tax, $3 40/100 on 17 members

at 20 cents each for the month of December, 1936;

Initiation Fees, $ /100; Reinstatement Fees, $ /100; Assessment, $ /100;

Charter and Outfit, $ /100; Supplies, $ /100; Misc., $ /100.

*Membership cards are mailed to the Financial Secretary of each Local, as per instructions printed on reverse side of this receipt. We herewith acknowledge, with thanks, your remittance. Fraternally yours,*

GEORGE J RICHARDSON,
Sec.-Treas., International Association of Fire Fighters.

TOTAL, $3 40/100 By

**IAFF MONTHLY DUES RECEIPT, FEBRUARY 1937.** Ashtabula's IAFF Local 165 received this receipt from the IAFF (based in Washington, D.C.) for its December 1936 monthly dues payment. Today 270,000 members strong, the IAFF encompasses 2,900 union locals serving over 3,500 U.S. and Canadian communities, including 238 Ohio locals. Ashtabula's IAFF Local 165, Ohio's seventh oldest, is predated by locals in Hamilton, Cincinnati, Columbus, Toledo, Cleveland, and Dayton. (Courtesy AFD Archive.)

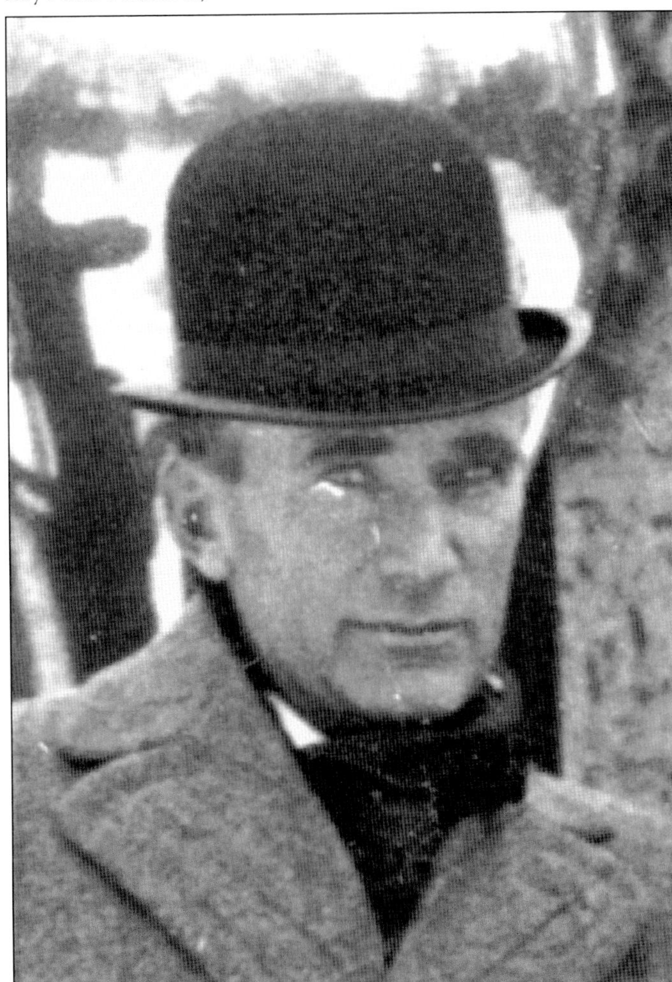

**CHARLES L. DEVEREAUX.** Pennsylvania oil field worker Charles Devereaux (1873–1943) relocated to Ashtabula Harbor in 1893, logging 15 years as a Pittsburgh Coal Company dock engineer and three years with Lake Shore Bridge Company. A 10-year Topky Fire volunteer, foreman Devereaux became a charter professional Ashtabula firefighter in October 1911. Promoted to captain in 1912, Devereaux became Ashtabula's first civil service fire chief in 1939, serving until his death in 1943. (Courtesy AFD Archive.)

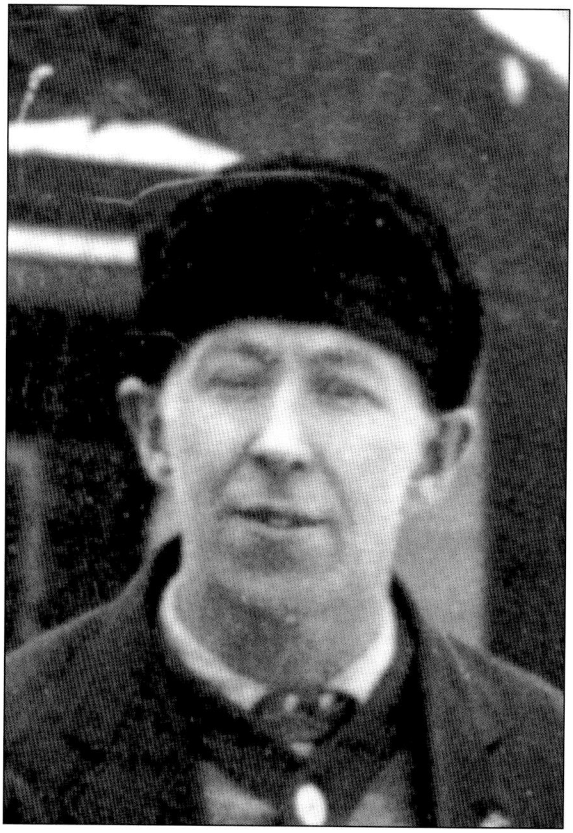

**FRED D. ALEXANDER.** Pennsylvania Railroad worker and Spanish-American War veteran Fred D. Alexander (1877–1944) devoted 40 years to city firefighting. Logging eight years with Ashtabula Harbor's volunteer Raser Hose Company (1903–1911), Alexander was a charter member of Ashtabula's paid professional fire department, promoted to lieutenant in 1919 and captain in 1930. Alexander served a seven-month tenure as fire chief between his October 1943 appointment and May 1944 death. (Courtesy AFD Archive.)

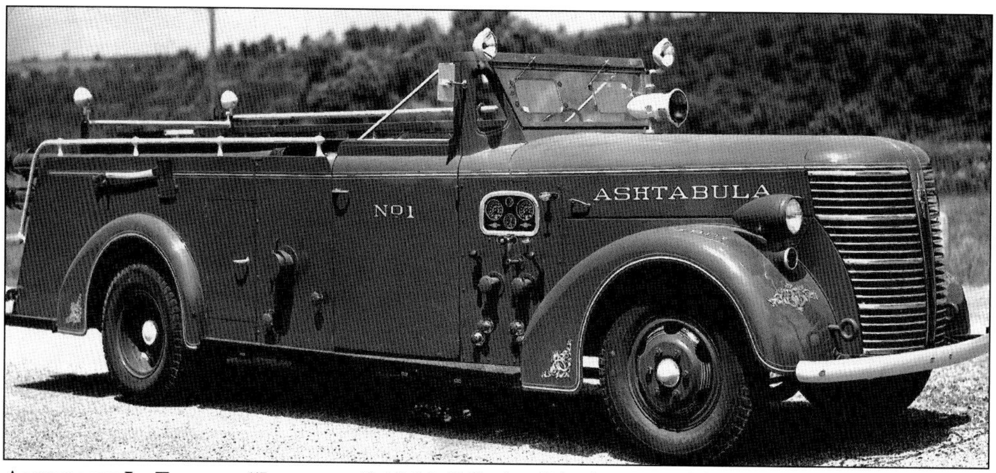


**AMERICAN-LAFRANCE "INVADER," 1944.** Billed as "the latest and finest of its kind," Ashtabula purchased this $9,160 American-LaFrance "Invader" pump and hose car to replace its *c.* 1911 Seagrave hose and chemical truck. Ordered by city manager W. H. Flower (1939–1944) and fire chief Fred D. Alexander (1943–1944), purchase of the 750-gallon-per-minute pumper was approved by the War Production Board and funded by a one mill, two-year tax levy. (Courtesy Ashtabula Star Beacon.)

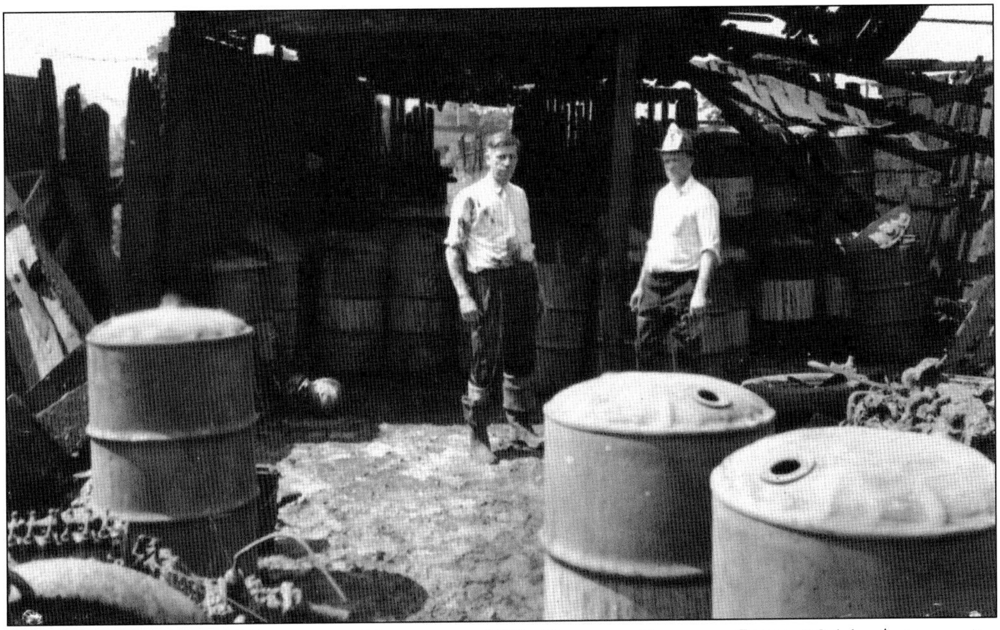

**ASHTABULA AUTO WRECKING COMPANY, 751 CENTER STREET, JULY 1944.** Acting city fire chief Capt. Adolph T. Rohl (left) and Lt. Arthur C. Larson sift through the remains of a downtown fire at Louis Silverberg's Ashtabula Auto Wrecking Company. A 30-year department veteran between 1916 and 1946, Larson was promoted to driver in 1919, lieutenant in 1930, and captain in 1945. (Courtesy AFD Archive.)

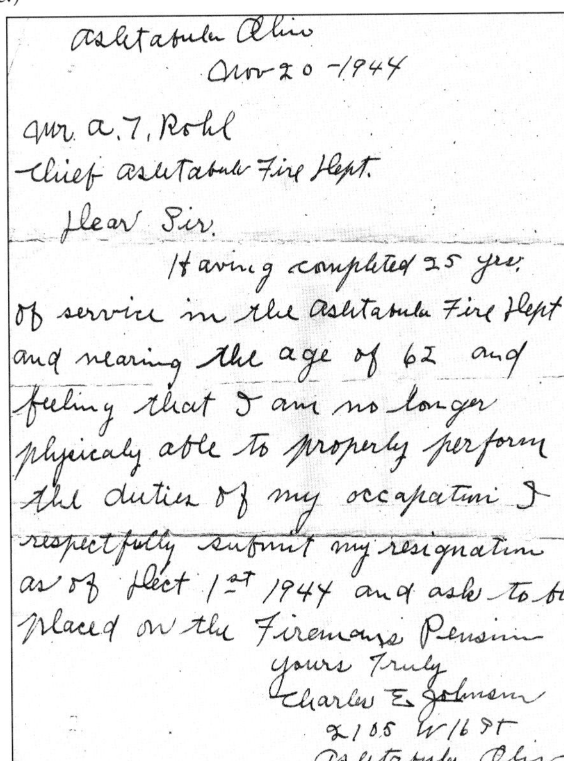

Ashtabula Ohio
Nov 20 - 1944

Mr. A. T. Rohl
Chief Ashtabula Fire Dept.
Dear Sir.
Having completed 25 yrs of service in the Ashtabula Fire Dept and nearing the age of 62 and feeling that I am no longer physicaly able to properly perform the duties of my occapation I respectfully submit my resignation as of Dect 1st 1944 and ask to be placed on the Firemans Pension
Yours Truly
Charles E Johnson
2105 W 16 St
Ashtabula Ohio

**RETIREMENT RESIGNATION, NOVEMBER 20, 1944.** Among the benefits of IAFF membership was the availability of the union's Fireman's Pension. Here 25-year city firefighter Charles E. Johnson, 62, tenders his handwritten resignation to acting city fire chief Rohl, noting he feels he is "no longer physically able to properly perform the duties of my occupation." (Courtesy AFD Archive.)

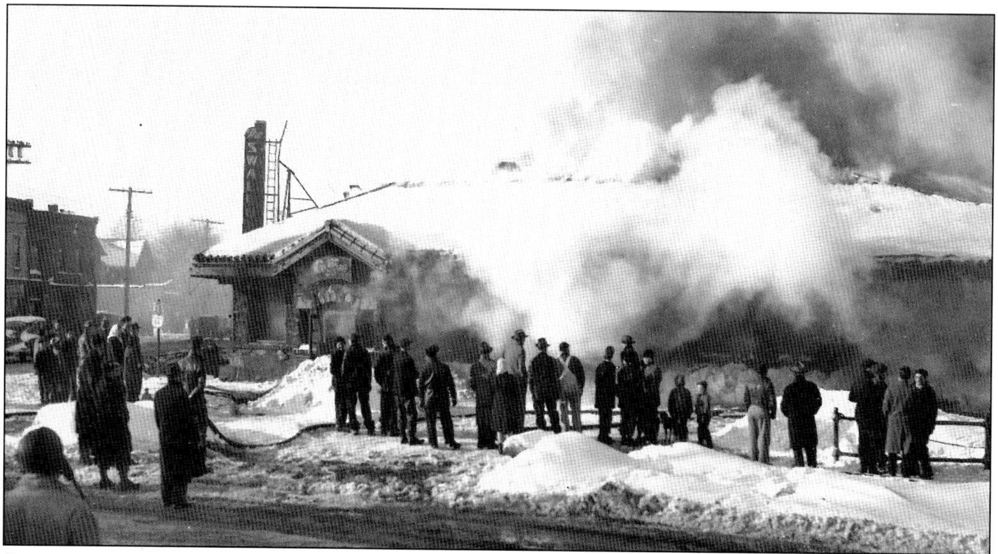

**SWALLOWS RESTAURANT, 1107 PROSPECT AVENUE—DECEMBER 29, 1944.** This dramatic blaze gutted five adjacent West Prospect Avenue businesses, including the Cook family's popular Swallows Restaurant (1939–1966). The nationally famous Swallows, opened in 1939 by former Ashtabula mayor, prominent local businessman, and "Sage of Bunker Hill" H. D. Cook, brought Ashtabula residents a parade of well-known entertainment stars including Duke Ellington, Count Basie, Woody Herman, Stan Kenton, Lionel Hampton, and Guy Lombardo and his Royal Canadians. Popular with local residents and tourists alike, the popular west side restaurant also hosted celebrity diners, including Hollywood actors Bette Davis and Warner Olin and numerous sitting Ohio governors and U.S. senators. A center of Ashtabula social life, Swallows hosted thousands of wedding receptions, proms, formal dances, Bar Mitzvahs, and association and club meetings. (Courtesy AFD Archive.)

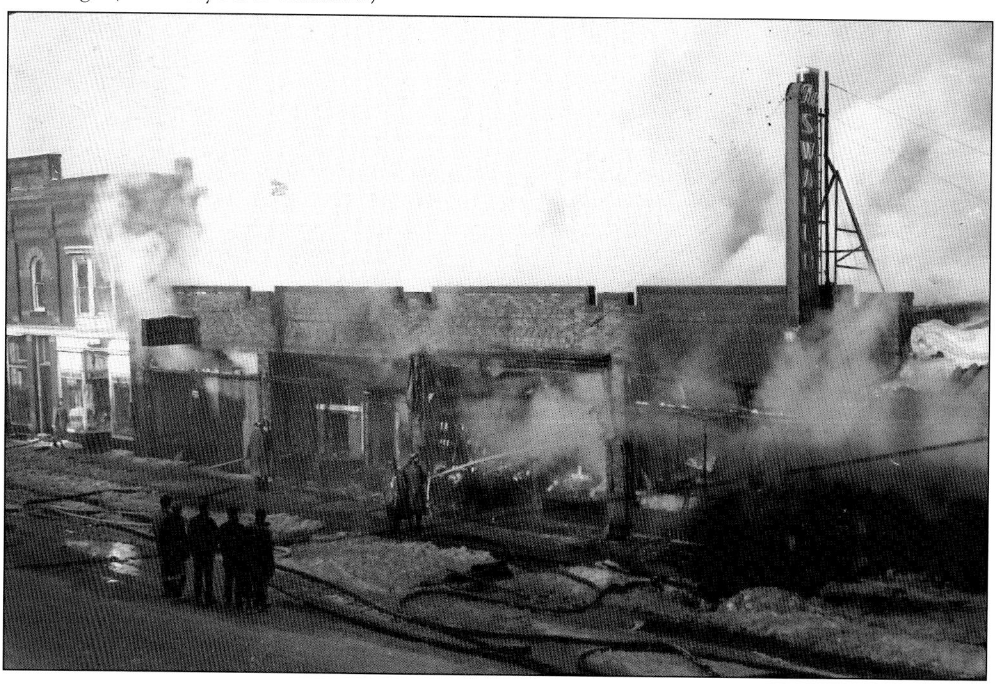

# *Four*

# The "Smokey" Rohl Era

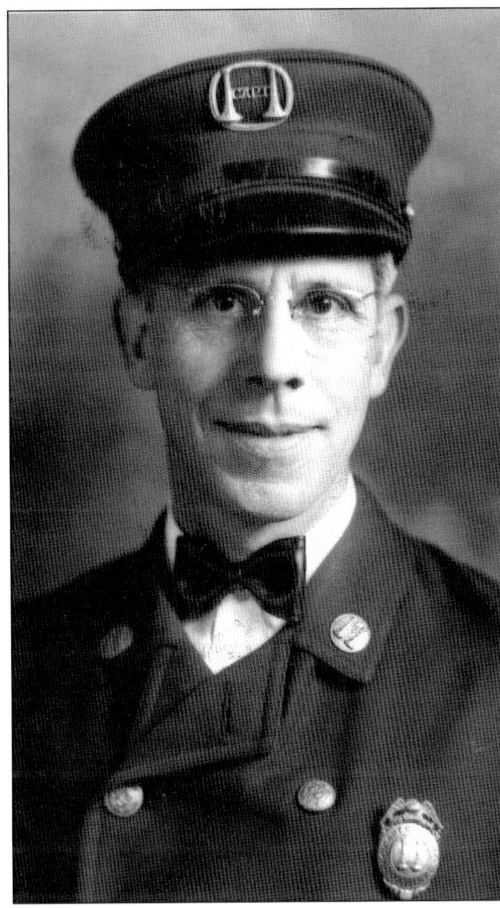

**Adolph T. Rohl Appointed Fire Chief, January 25, 1945.** Following the unexpected May 28, 1944, death of fire chief and longtime department veteran Fred D. Alexander, city manager W. H. Flower tapped veteran Ashtabula firefighter Adolph T. "Smokey" Rohl to serve as acting fire chief. In January 1945, following Rohl's successful completion of the civil service examination, acting Ashtabula city manager Fred R. Hogue announced Rohl's appointment as fire chief, a role he would fill for 18 years. A longtime Uptown Fire Station fixture, Rohl joined the department as a hoseman in January 1913. Promoted to driver in 1917 and lieutenant in 1930, Rohl rose to the rank of captain in 1939. With Rohl's appointment as fire chief, 29-year department veteran Lt. Arthur C. Larson became Uptown's new captain. (Courtesy Ashtabula Star Beacon.)

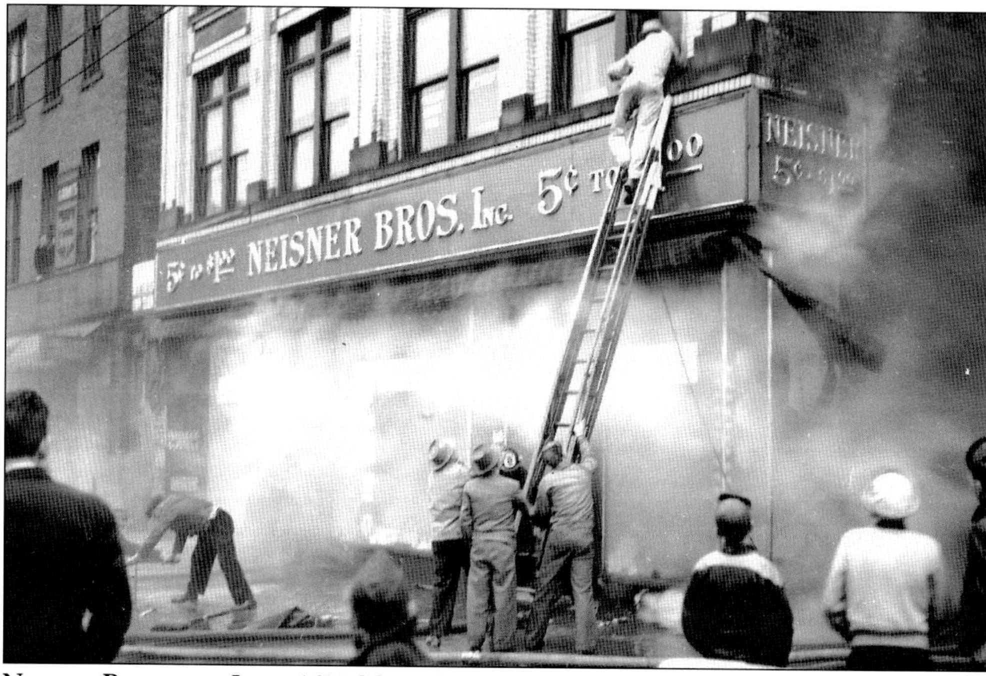


**Neisner Brothers, Inc., 4623 Main Avenue—May 4, 1946.** A late Saturday afternoon blaze at the popular Neisner Brothers five-and-dime variety store in downtown's 1870s Schaffner Block sent nearly 60 patrons scurrying for the exits. The $100,000 blaze, sparked by a furnace room explosion, burned out Rochester, New York–based Neisner's (1910–1978), praised as "one of the leading stores of its kind in the area." The fast-spreading and stubborn eight-hour fire was fought by 100 firefighters from Ashtabula, North Kingsville, Jefferson, Conneaut, Andover, Austinburg, and Rock Creek. Neisner's returned to downtown Ashtabula in May 1949, unveiling a new state-of-the-art store designed by Ashtabula architect C. V. Martin at 4627 Main Avenue. (Courtesy AFD Archive.)

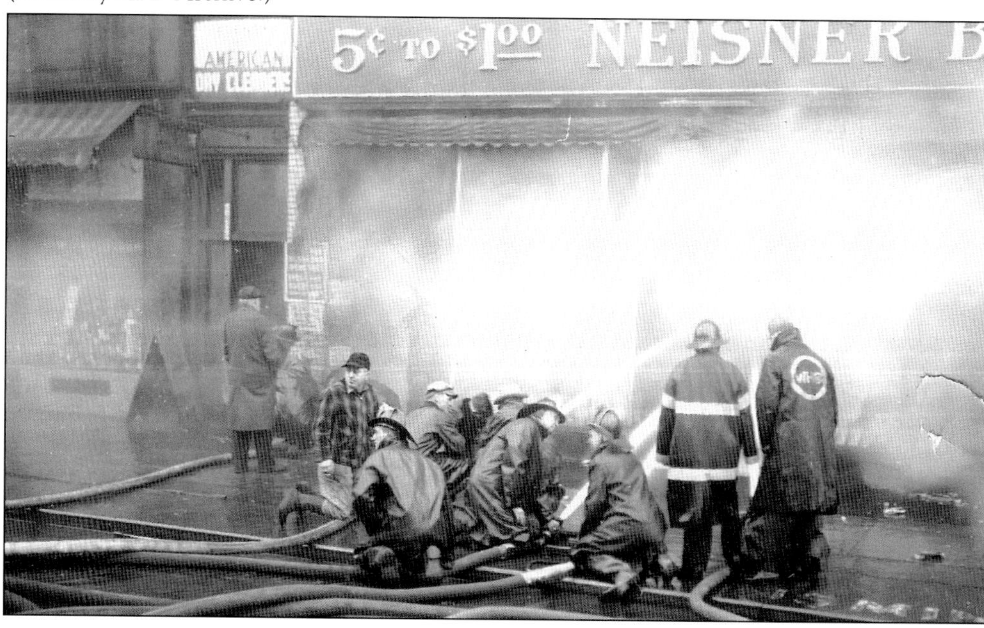

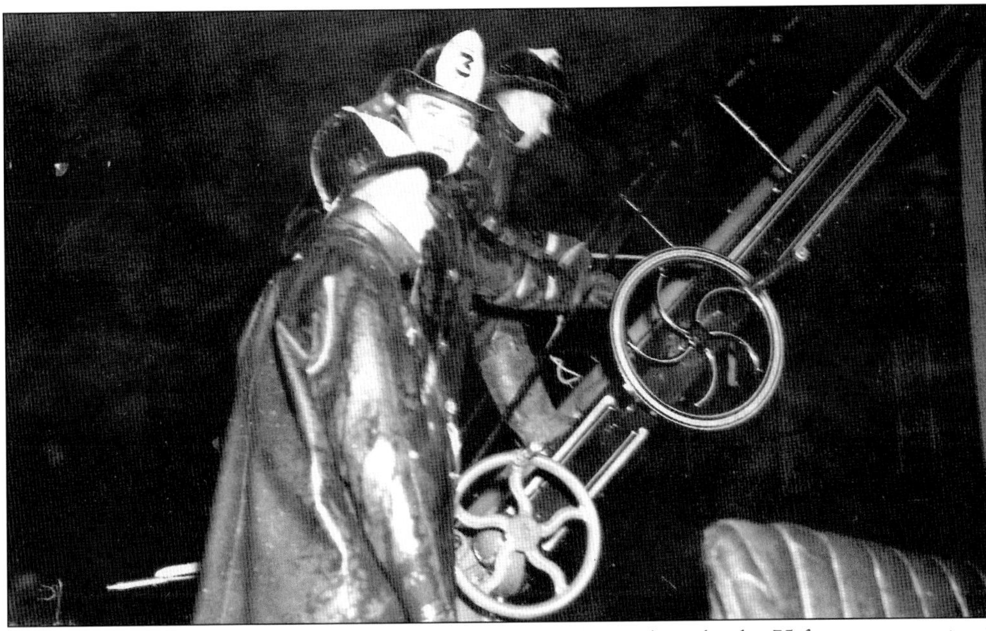

**RAISING THE AERIAL LADDER.** Capt. Thomas P. Kane (center) works the 75-foot spring-assisted ladder on Ashtabula's *c.* 1930 Seagrave hook and ladder truck, which offered 305 feet of ladders between the aerial ladder and the 10 ground ladders carried on the truck frame. The bottom wheel on the aerial ladder governed rotation, while the top wheel extended the ladder. Twin hand cranks on the other side, hidden from view, adjusted elevation. (Courtesy AFD Archive.)

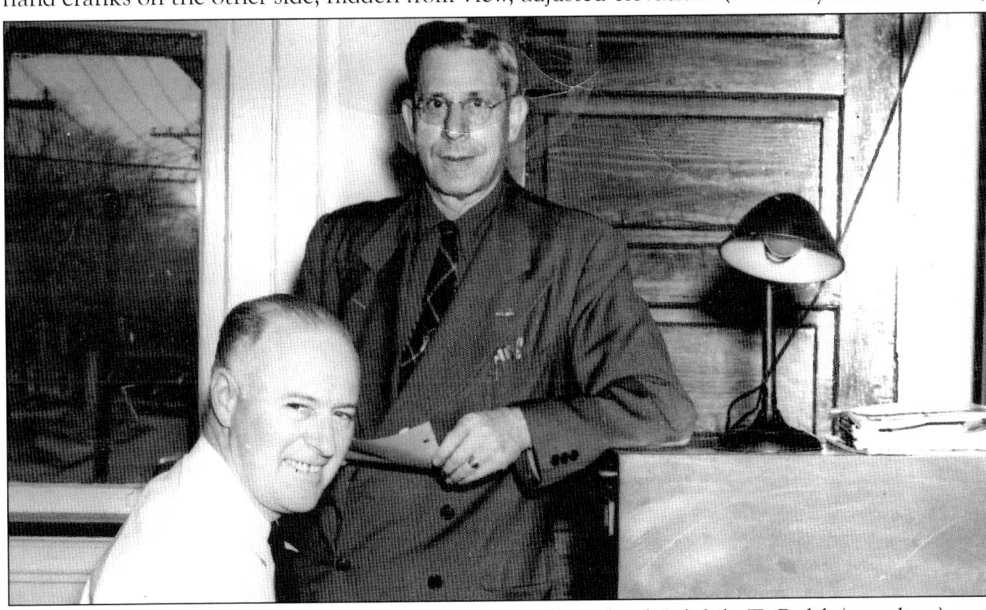

**HEADQUARTERS OFFICE, MID- TO LATE 1940S.** Fire chief Adolph T. Rohl (standing) and Capt. Thomas P. Kane are pictured in the Ashtabula Fire Department's second-floor headquarters office at the Uptown Fire Station, 4758 Park Avenue. A longtime department veteran, Kane joined the department as hoseman in April 1930. Promoted to engineer in 1939, lieutenant in 1943, and captain in 1945, Kane retired in April 1971 after a 41-year career. (Courtesy AFD Archive.)

## SECTION XI

### Station Houses

1. The grounds and sidewalks around the station house and the front of the structure itself shall be kept clean and orderly at all times.
2. Interior housekeeping shall be constantly maintained at a high level.
3. Beds shall be made up neat and clean by 9:00 A. M. each day except on days when bedding is changed or aired, when the beds shall be made up by noon.
4. No member of the Division shall occupy his station-house bed between 8:30 A. M. and 8:00 P. M. except in cases of illness or fatigue from duty the preceding night.
5. Members of the Division shall not sleep, or recline in such a position as to indicate sleeping, on the apparatus or in the apparatus room.
6. Members of the Division shall not be disturbed unnecessarily when sleeping in dormitories during the prescribed hours.

**DIVISION OF FIRE—RULES AND REGULATIONS, APRIL 1946.** Under city manager R. M. Hoisington and new fire chief Adolf T. Rohl, the city issued 12-chapter, 27-page "Rules and Regulations" booklets to Ashtabula firefighters in the interest of creating "a well organized unit characterized by a spirit of cooperation, harmony, and discipline among its members." This book once belonged to Lt. John C. Johnson (1928–1955). (Courtesy AFD Archive.)

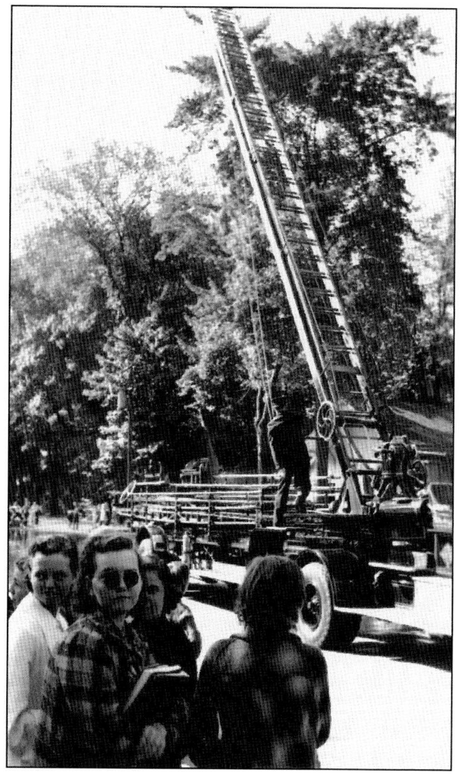

**SEAGRAVE HOOK AND LADDER TRUCK, LATE 1940S.** A crowd of high school girls watch Ashtabula firefighters raise the 75-foot aerial ladder on their Seagrave hook and ladder truck (1930–1966). Ashtabula Township resident Harold Osten, the Seagrave's longtime owner, donated the aerial truck to the non-profit Fire Historical Society of Ashtabula in 2003. The society, headed by 13-year firefighter Capt. Jeff Gianantonio, hopes to raise more than $15,000 for the Seagrave's restoration. (Courtesy AFD Archive.)

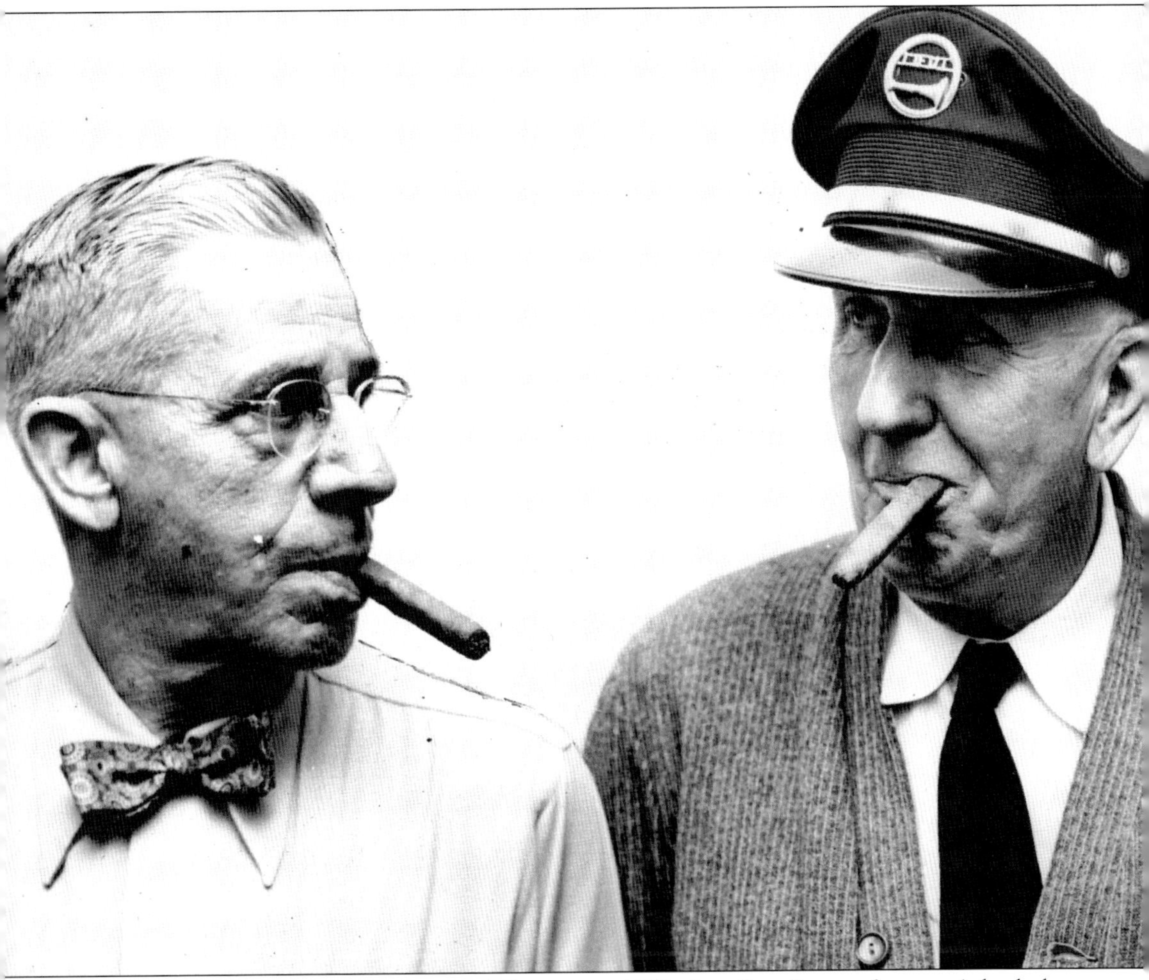

**"HAVE HAVANA—WILL FIGHT FIRES," FEBRUARY 1950.** In a *Star Beacon* feature, Ashtabula fire chief Adolph T. "Smokey" Rohl (left) and Lt. Henry C. Froid were said to have earned their lighthearted motto through their "trick" use of unlit cigars to filter breathable air at fires: "The chief explains he's not smoking the cigar. He's 'breathing' through the cigar. He lets the cigar filter out smoke from the fire. Earlier this year, one fireman was felled by smoke when his . . . mask failed. He has taken up the use of the cigar filter trick. Lt. Harry Froid is another of the fire crew who can be seen chewing a cigar instead of eating smoke. It all goes to show that whenever there's a fireman and a cigar, there's not necessarily smoking." But in a 1963 retirement retrospective, Rohl admitted that their omnipresent trademark cigars had simply "provided good copy for reporters for years" as rookie newshounds fell for tall tales. Still, noted one admiring firefighter, "The chief can eat more smoke than any man in the department." (Courtesy Ashtabula Star Beacon.)

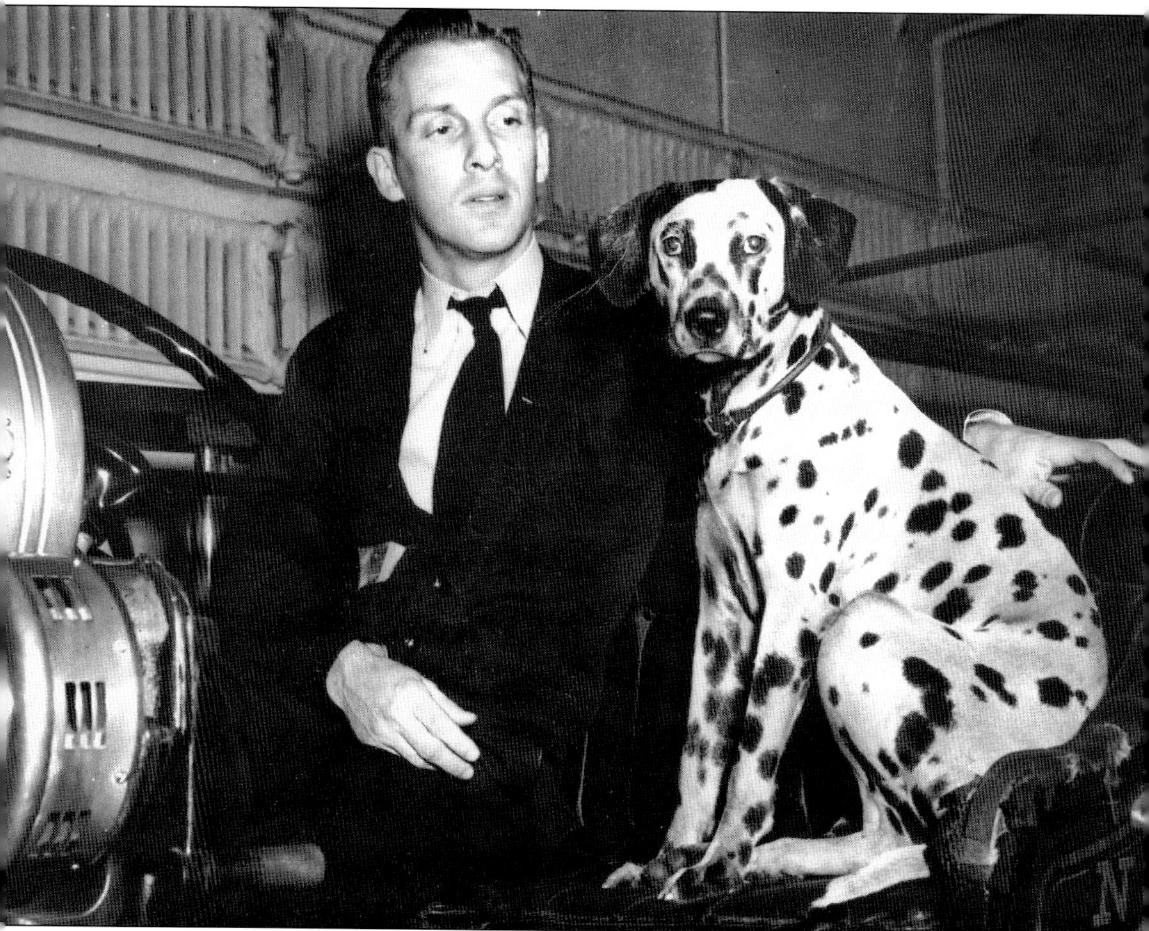

**Charles L. Mosier and Duchess, December 1950.** Fireman and future city fire chief Charles (Charlie) Mosier introduces 13-month-old "Duchess of MacMorland II," nicknamed "Lady," to Ashtabula residents in this *Star Beacon* photograph. Traditional fire station mascots, purebred Dalmatians were also once popular in stagecoach post houses. According to legend, the distinctive black-and-white dogs got their spots from axle grease drippings. Harbor Station firefighters, it was noted, were planning to outfit Duchess with a collar and fire badge similar to that worn by the Conneaut Fire Department's Dalmatian, Smudge. Mosier and "affectionate and gentle" Lady are pictured on the seat of the Harbor Fire Station's *c.* 1922 American-LaFrance Foamite pumper, Lady's bed when the truck was not in action. A Great Lakes freighter cook took a shine to Duchess during a prolonged Port of Ashtabula call, bringing her "bones and choice tidbits from the galley" on a daily basis and making a special trip to the station by taxi to "bid her farewell." (Photograph Ashtabula Star Beacon; courtesy AFD Archive.)

**NEW EMERGENCY MEDICAL EQUIPMENT, AUGUST 1951.** Harbor's Lt. Edward St. Ledger (1930–1952) showcases Ashtabula's newly delivered Emerson resuscitator, inhalator, and aspirator system, based at the Harbor Fire Station for quick emergency medical deployment at Walnut Beach and Lakeshore Park. Trained in medical rescue beginning in 1948, Ashtabula firefighters were certified in CPR by 1968, Red Cross first aid by 1972, and EMT in the mid- to late 1970s. (Courtesy Ashtabula Star Beacon.)

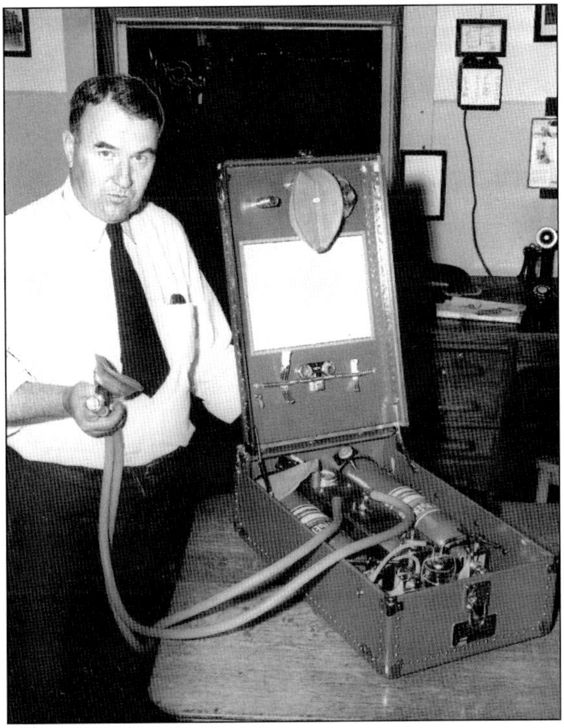

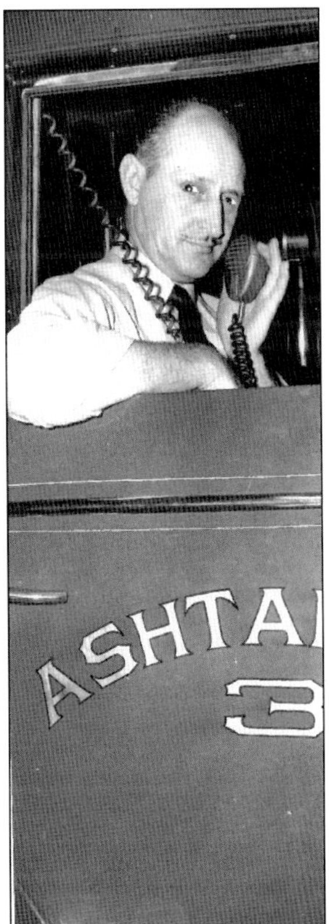


**NEW TWO-WAY RADIO, AUGUST 1951.** Sitting in Truck 3, Uptown firefighter Lawrence Byrde (L. B.) Thomas demonstrates the use of the Ashtabula Fire Department's new two-way radio. The unit was part of the police department's new high frequency FM system. As part of the new communication system, the city also installed a new central emergency telephone system for the fire and police departments. (Courtesy Ashtabula Star Beacon.)

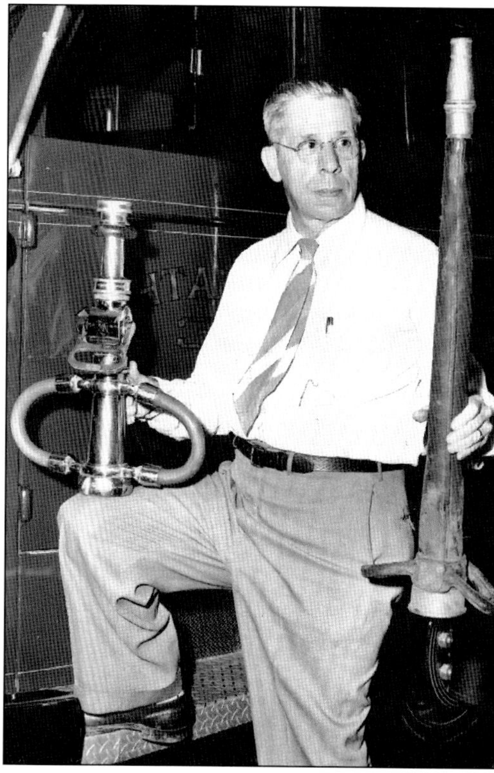

**ROHL MARKS 40TH ANNIVERSARY, 1953.** Exhibiting past and present firefighting technology on his 40th anniversary with the Ashtabula Fire Department, fire chief Adolph T. "Smokey" Rohl showcases a state-of-the-art hose nozzle (left) with its 1889 predecessor, used by Ashtabula's volunteer Protection Fire Company. Joining the department in 1913, Rohl saw numerous firefighting innovations including fog nozzles, carbon dioxide extinguishers, better hose, and increasingly sophisticated fire trucks. (Courtesy Ashtabula Star Beacon.)

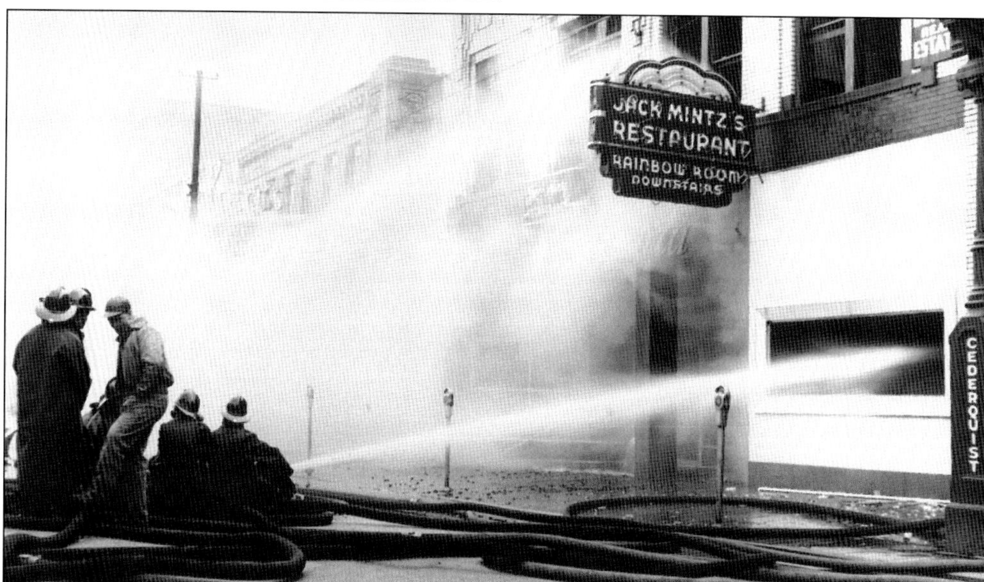


**JACK MINTZ RESTAURANT AND LOUNGE, 4644 MAIN AVENUE—APRIL 26, 1954.** An extensive six-month-old remodeling job was destroyed in this $250,000 blaze at downtown's popular Jack Mintz Restaurant and Lounge. Ground and ladder crews were augmented by firefighters stationed atop Shea's Theater and the neighboring Crosby Block. Manning 10 fire lines were 60 firefighters from the Ashtabula, North Kingsville, and Kingsville fire departments. (Photograph Ashtabula Star Beacon; courtesy AFD Archive.)

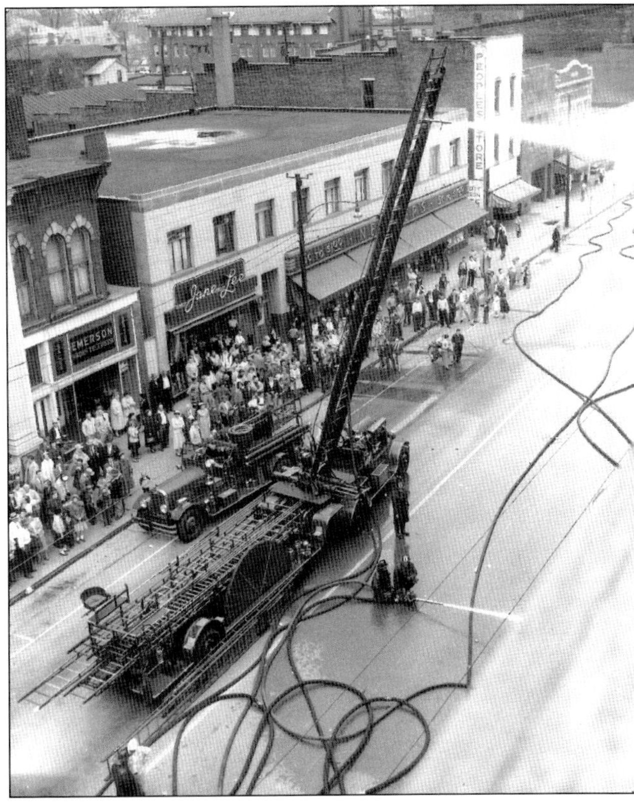


**Jack Mintz Restaurant and Lounge, 4644 Main Avenue—April 26, 1954.** Seen from the six-story Hotel Ashtabula, the Ashtabula Fire Department's 75-foot c. 1930 Seagrave aerial truck works a $250,000 fire at downtown's popular Jack Mintz Restaurant and Lounge. The stubborn seven-and-a-half-hour early morning blaze drew a crowd of over 1,000 curious onlookers and a live news remote by Ashtabula's WICA-AM 970. (Photograph Ashtabula Star Beacon; courtesy AFD Archive.)

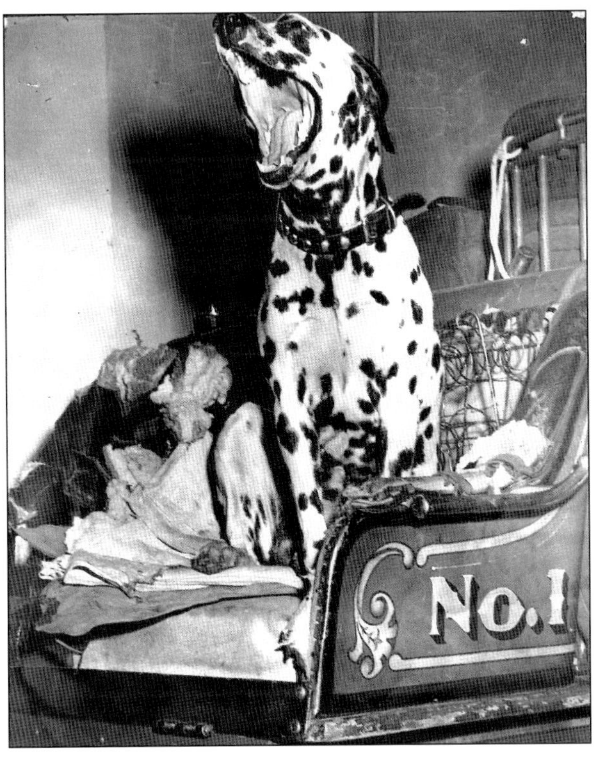


**Time for a New Truck, 1955.** Harbor's popular Duchess of MacMorland II sits on her uncomfortable, decrepit bed seat atop Ashtabula's 1922 American-LaFrance pumper. Duchess' plight shamed Ashtabula into replacing its 33-year-old truck with a new 1,000-gallon-per-minute Seagrave pumper. A late 1950s *Star Beacon* photograph of Duchess huddled against a stove at the drafty 1880s Harbor Fire Station spurred construction of a new facility. (Photograph Ashtabula Star Beacon; courtesy AFD Archive.)

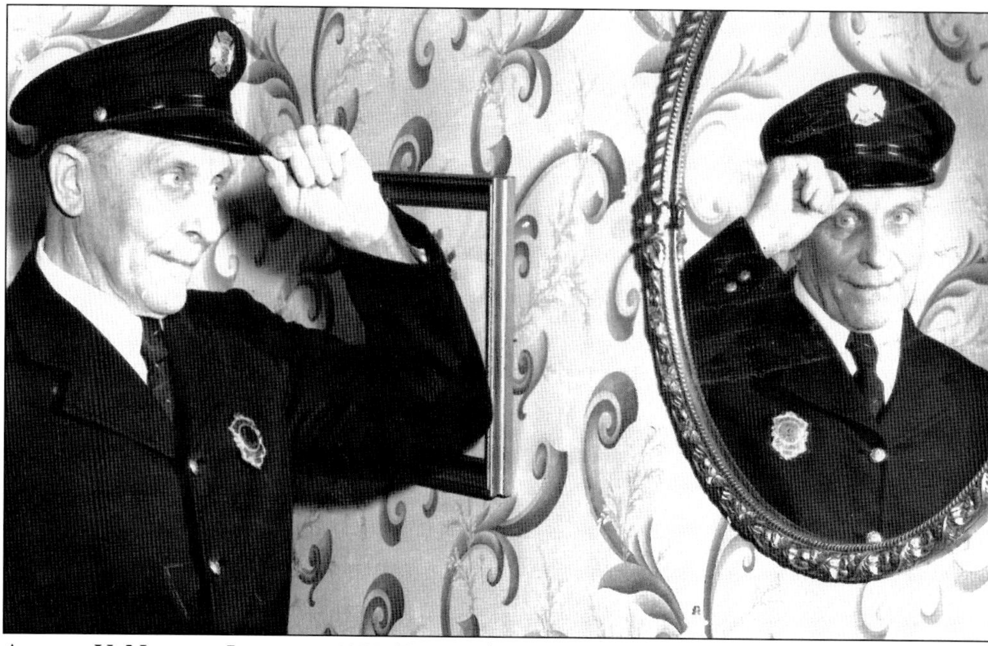

**ALBERT V. NELSON, JANUARY 1957.** For nearly 43 years between May 1914 and January 1957, hoseman Albert Nelson was a Harbor Fire Station fixture. Inside his West Fourth Street home, Nelson prepares for his final shift. Appointed by mayor Dr. I. H. Pardee at an initial pay rate of $66 per week, Nelson missed only one day's work over his career after suffering facial burns. (Photograph Ashtabula Star Beacon; courtesy AFD Archive.)

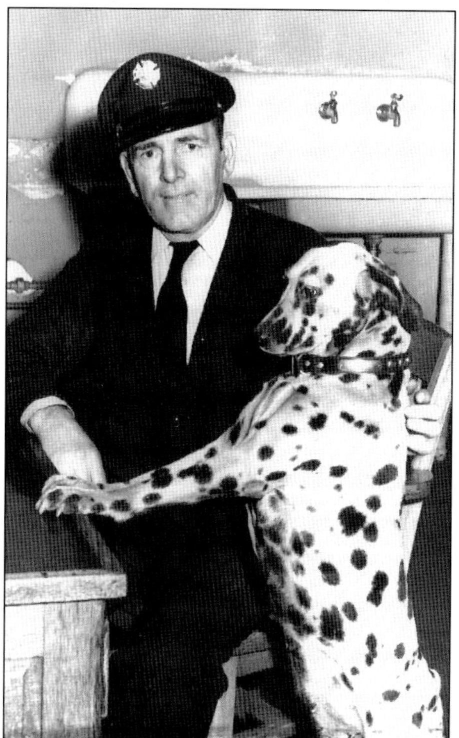

**MAURICE P. SHERRY AND DUCHESS, 1958.** Billed as "the Harbor's most photographed dog," Harbor Fire Station's beloved Duchess of MacMorland II (1949–1958) is pictured with firefighter Maurice Sherry, who had purchased the thoroughbred Dalmatian in Cleveland. Wrote the *Star-Beacon* of Duchess in 1950, "Though officially only a duchess in name, all the Harbor firemen readily admit she is 'queen of the station.'" (Photograph Ashtabula Star Beacon; courtesy AFD Archive.)

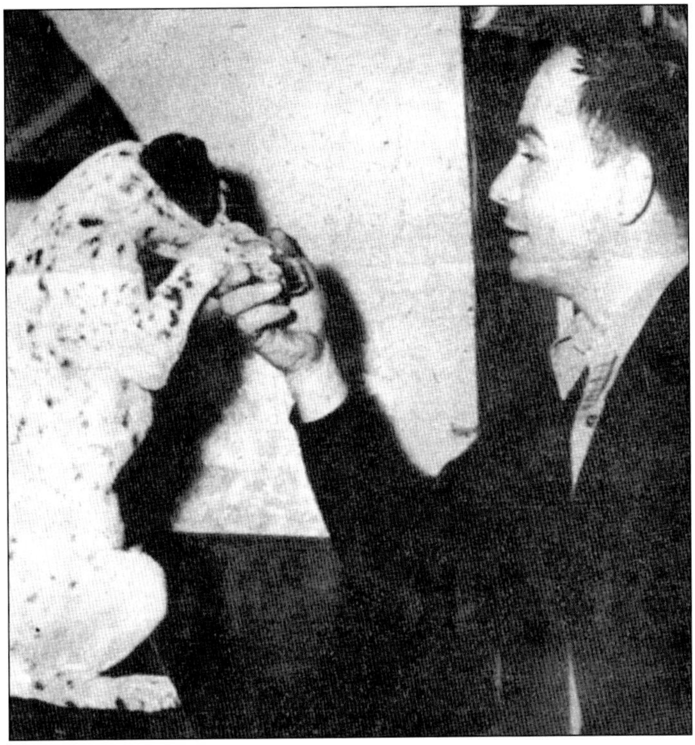

**"GEE WHIZ! I'M A FIREMAN NOW," 1958.** Following the January 1958 beating death of Duchess by an unknown Bridge Street assailant, Orwell resident Jack Smiley presented grieving Harbor firefighters with three-month-old thoroughbred Dalmation "Smokey," seen meeting firefighter and World War II veteran Octavian "Barney" Pugliese (1953–1982). Pugliese, later a captain and training officer, recalls Smokey as "a very nice, friendly, good-natured dog." (Courtesy Ashtabula Star Beacon.)

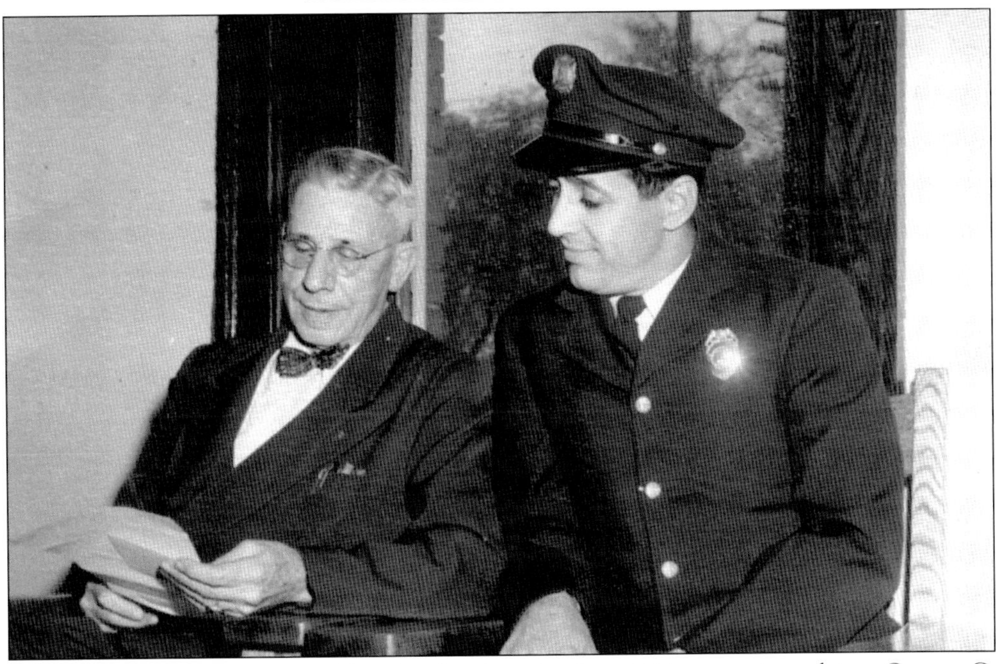

**KEEPING THE BOOKS, 1958.** Shortly after his October 1957 promotion, driver George C. Matrisciano (right) and 1945–1963 fire chief Adolph T. "Smokey" Rohl examine department record books. A stickler for meticulous record keeping, Rohl took pride in keeping department paperwork up-to-date and in order. Joining the department in 1950, 28-year veteran Matrisciano retired as a captain in 1978. (Photograph Ashtabula Star Beacon; courtesy AFD Archive.)

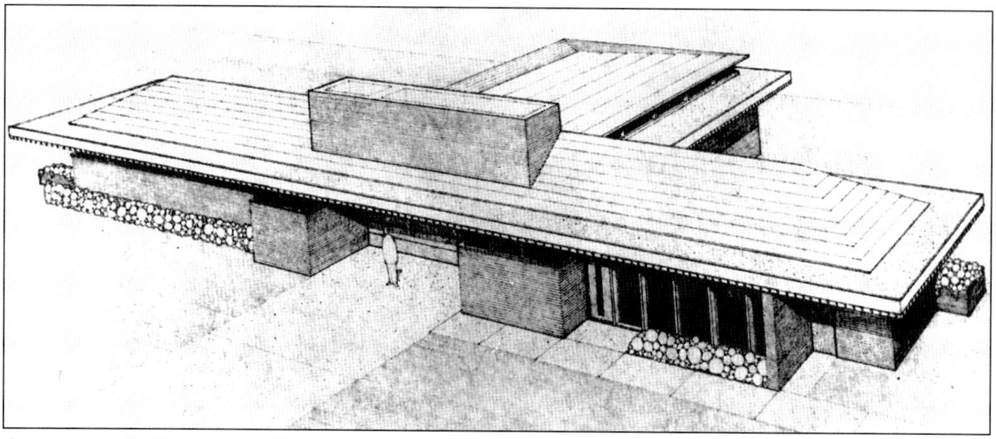

**ARCHITECT'S DRAWING, HARBOR FIRE STATION, 1958.** In 1955, city council and city manager Ned V. Collander approved construction of a new Harbor Fire Station to replace the city's 1880s Bridge Street station. An undeveloped 3.8-acre site at 1926 West Ninth Street was purchased from Frank A. and Ann Nappi in 1958, and construction began on the residential-styled station, designed by Ashtabula architect George Anderson. (Courtesy Ashtabula Star Beacon.)

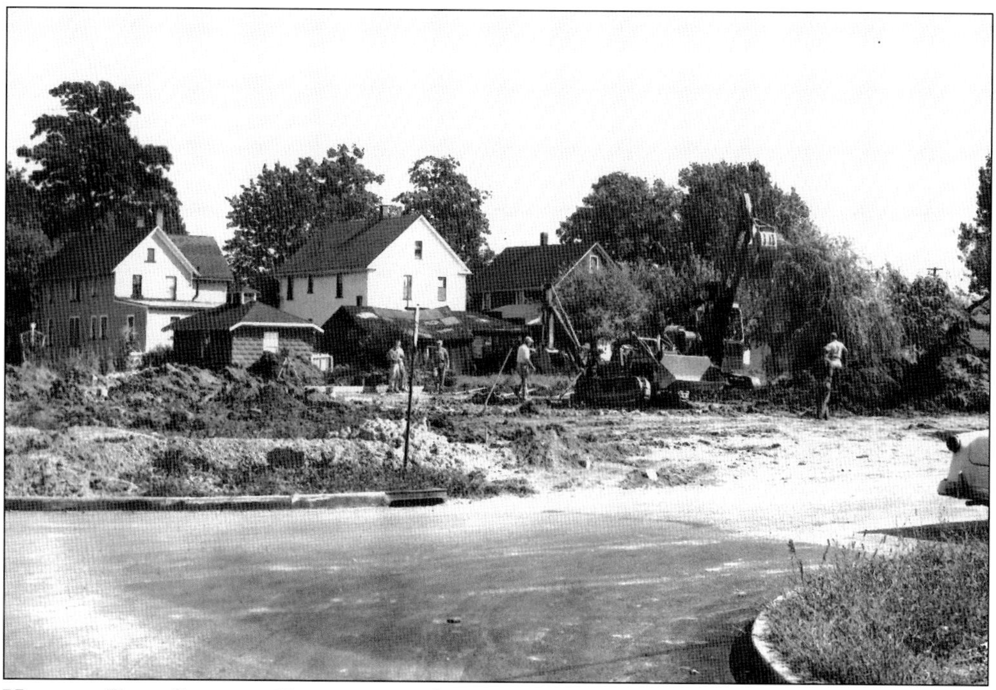

**HARBOR FIRE STATION EXCAVATION, SEPTEMBER 1958.** Excavation work begins for the construction of Ashtabula's new $85,780 Harbor Fire Station, located on the southeast corner of West Ninth Street and Michigan Avenue. Jefferson's Joseph Nizen was project contractor. Funding for the 3,300-square-foot station came from the city's Permanent Improvement Fund, established in 1947 for projects with an anticipated lifespan of five years or more. (Courtesy Ashtabula Star Beacon.)

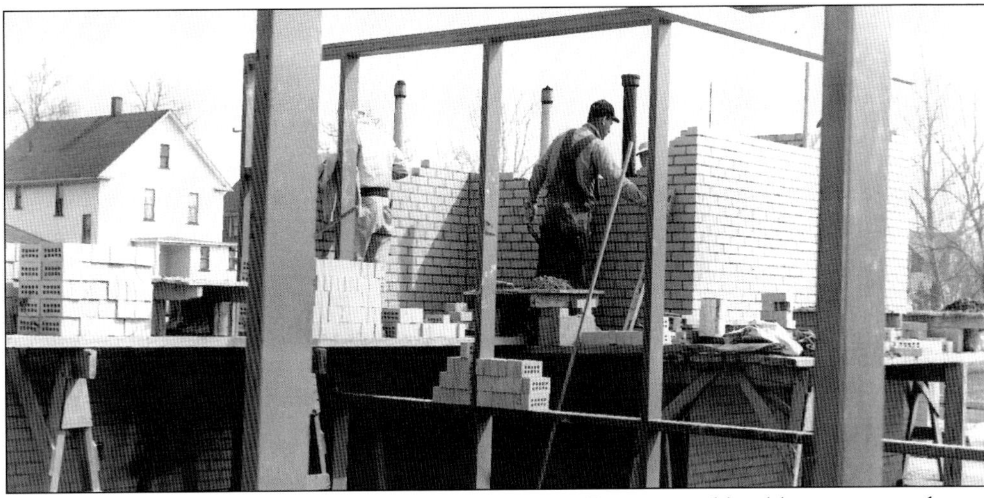

**HARBOR FIRE STATION CONSTRUCTION, APRIL 1959.** Framing and bricklaying gets underway on the city's new Harbor Fire Station. Winter weather, coupled with tight supplies of construction materials, delayed the project several times, pushing its completion back from July 1959 to mid-October. At its dedication, the *Star Beacon* feted the Harbor Fire Station as "a new symbol of progress to gladden the hearts of citizens." (Courtesy Ashtabula Star Beacon.)

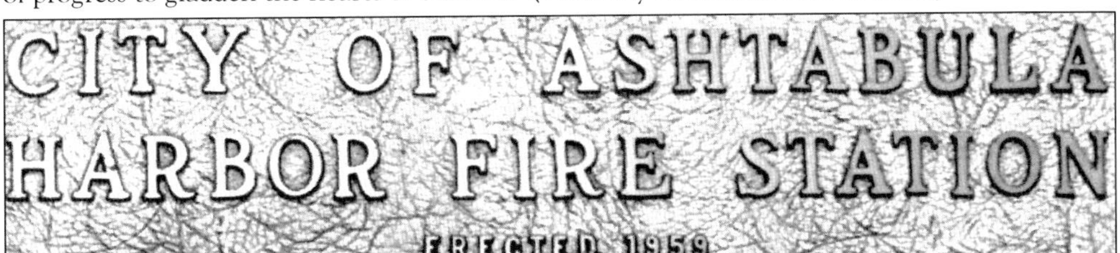


**HARBOR FIRE STATION DEDICATION PLAQUE, 1959.** Ashtabula's aging 1880s Harbor Fire Station on Bridge Street was replaced in 1959 by a state-of-the-art, low-rise, residential-styled station at West Ninth Street and Thayer Avenue. Designed by architect George Anderson and built by Jefferson contractor Joe Nizen, the station was praised by the *Star Beacon* as a facility that would create "a new spirit and improved morale" among Harbor firefighters. (Courtesy AFD Archive.)

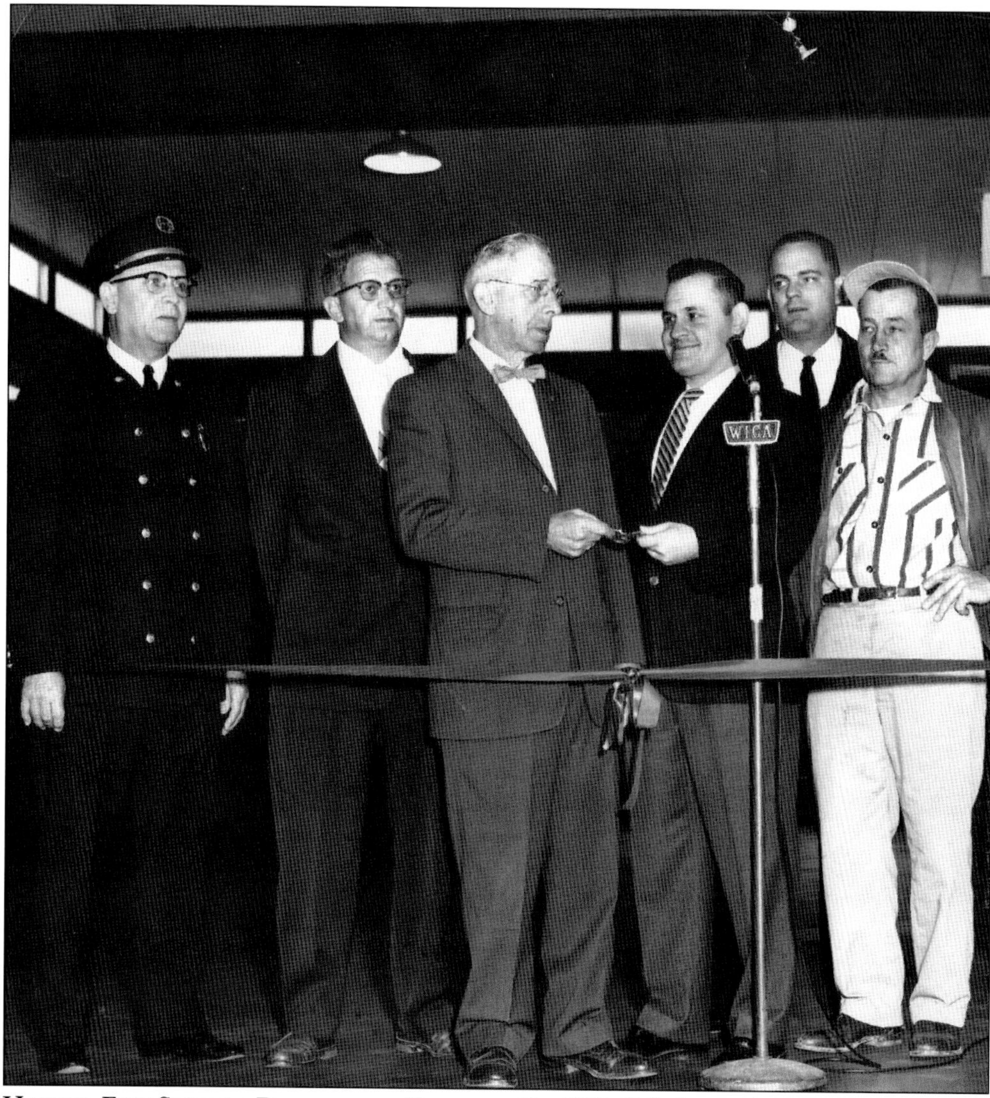


**HARBOR FIRE STATION DEDICATION, OCTOBER 12, 1959.** Billed as "a day to remember in the Harbor area and a day to mark in the history of Ashtabula," the four-year quest to replace the city's aging Harbor Fire Station on Bridge Street came to a successful close with the dedication of the Ashtabula Fire Department's state-of-the-art $85,780 Harbor Fire Station, 1926 West Ninth Street. The 3,300-square-foot low-rise brick structure, built to blend into its residential Harbor neighborhood, included a 12-bed dormitory, a two-truck equipment bay, and recreational and kitchen facilities. Dedication dignitaries included, from left to right, Harbor fire captain L. M. Thurber, city councilman Gerald Severino, fire chief Adolph T. Rohl, city manager C. D. Lambros, architect George Anderson, and Jefferson contractor Joseph Nizen. The ceremony, attended by 75 residents, was broadcast live on Ashtabula's WICA-AM 970. Here Lambros presents Rohl with scissors to perform the ceremonial ribbon cutting, noting Rohl "deserved the privilege" as the person most responsible for making the new station a reality. (Courtesy Ashtabula Star Beacon.)

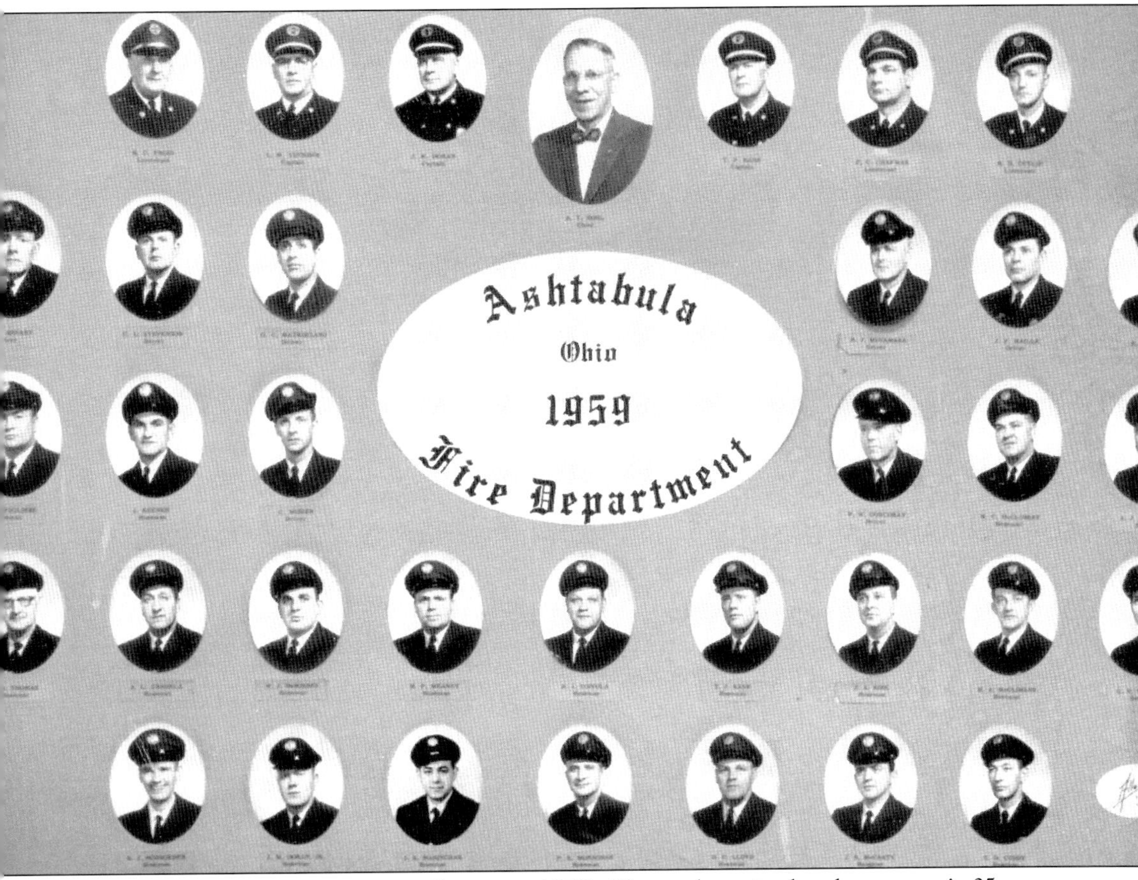


**ASHTABULA FIRE DEPARTMENT ROSTER, 1959.** Shown here is the department's 35-man, two-station 1959 roster. From left to right are (first row) hosemen E. J. Schroeder, J. M. Doran Jr., J. R. Marinchak, P. E. Monaghan, D. C. Lloyd, J. R. McCarty, and G. D. Cuddy; (second row) hosemen L. B. Thomas, A. L. Candela, W. J. DeSorbes, M. P. Meaney, W. I. Toivola, T. J. Kane, J. A. Kirk, H. A. McClimans, and G. E. Van Allen; (third row) hosemen O. R. Pugliese and J. Keenan, drivers C. L. Mosier and P. W. Corcoran, and hosemen W. C. McCluskey and A. J. Ucciferri; (fourth row) Drivers M. P. Sherry, C. L. Stevenson, G. C. Matrisciano, R. J. McNamara, J. F. Maille, and E. H. Aronhalt; (fifth row) Lt. H. C. Froid, Capt. L. M. Thurber, Capt. J. M. Doran, Fire Chief A. T. Rohl, Capt. T. P. Kane, Lt. P. C. Chapman, and Lt. R. D. Cutlip. (Photograph Allen Studios; courtesy AFD Archive.)

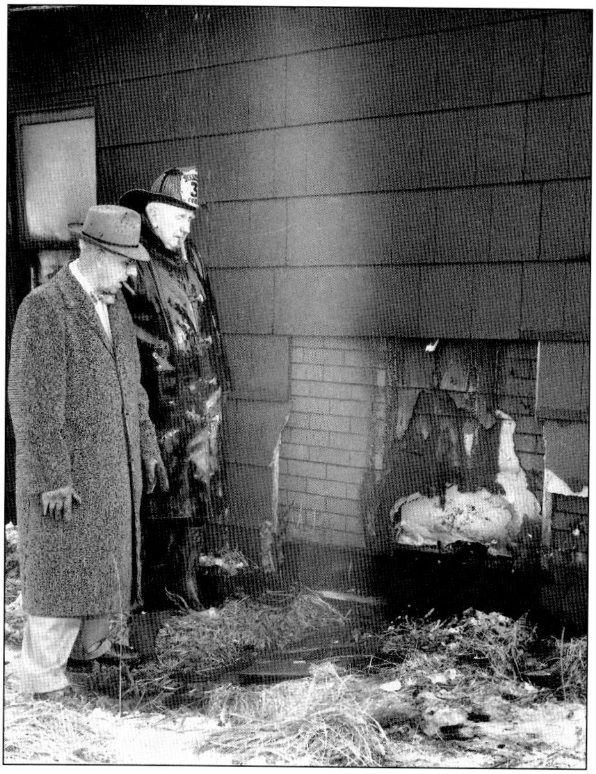

**"I Say, Holmes! Don't You Ever Light That Cigar With A Blow Torch Again!"** Behind the scenes, firefighters have a long tradition of good-naturedly teasing their colleagues with humorous creations, as evidenced by the inspired caption for this photograph of two veteran cigar-loving department colleagues, fire chief Adolf T. "Smokey" Rohl (left) and Lt. Harry C. Froid. Rohl was famed for his cigar-smoking antics. (Photograph Ashtabula Star Beacon; courtesy AFD Archive.)

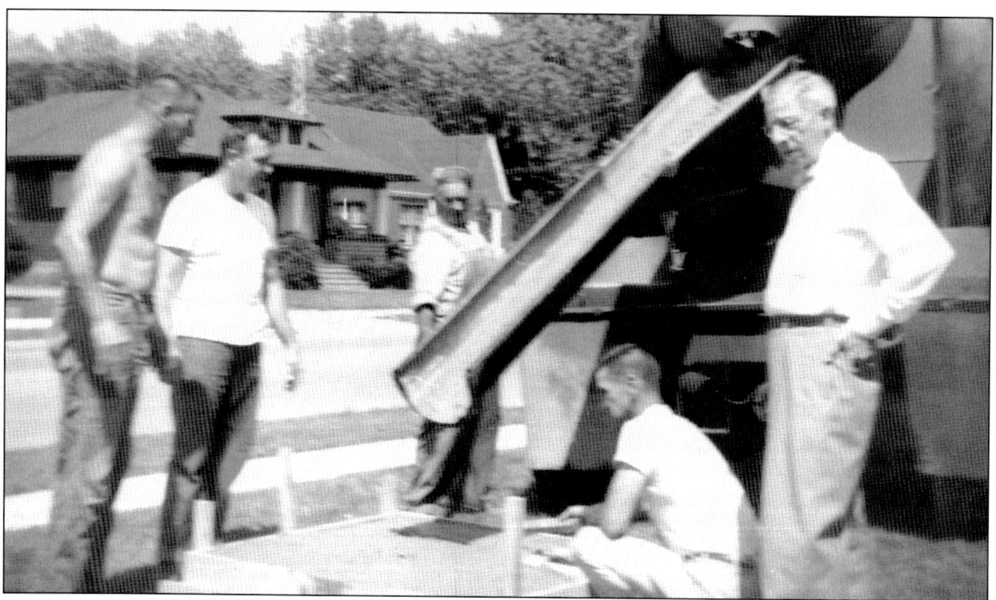

**Pouring Cement, July 1961.** As fire chief Rohl (right) looks on, a cement foundation is poured for the Harbor Fire Station's planned flagpole. City financial problems forced Harbor's January 1988 closure. Occupied by the Ashtabula Police Department from 1988 to 1996, the facility has been leased since 2005 to the Northeast Regional Response Group, a five-county hazardous materials unit serving Ashtabula, Lake, Trumbull, Geauga, and Portage Counties. (Photograph Dale G. Dowler Jr.; courtesy AFD Archive.)

**FLIP.** Uptown firefighters adopted Dalmatian Flip in the late 1940s. "Loyal fire-eater" Flip was famed for snuffing flaming matches with her paws, a trick firefighters vowed she had learned prior to her adoption. A friendly late-afternoon Nickel Plate Railroad dining car chef kept Flip supplied with bones, while firemen took turns taking her home for supper. An aged Flip was euthanized in the early to mid-1960s. (Courtesy Ashtabula Star Beacon.)

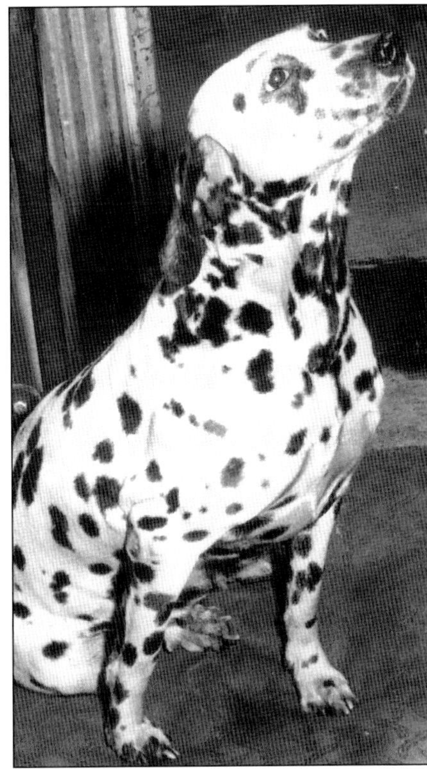

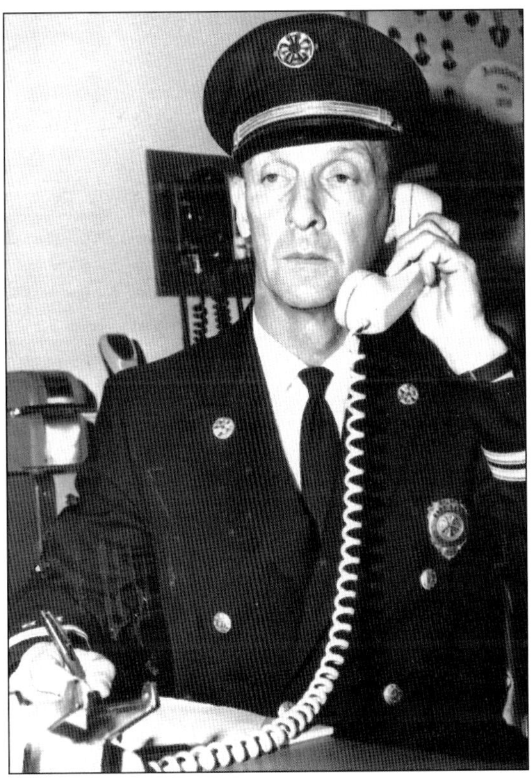

**LT. CHARLES L. MOSIER, 1961.** Numerous World War II veterans joined the Ashtabula Fire Department, including 1941 Ashtabula High School alumnus Charles Mosier (1947–1987). Serving in the Pacific theater with the U.S. Navy, Mosier served a 1942–1945 stint on the aircraft carrier USS *Antietam*, including two years as a fire brigade officer. Mosier's father, Emerson, and father-in-law, Merle Craisley, were both longtime Ashtabula police officers. (Courtesy Ashtabula Star Beacon.)

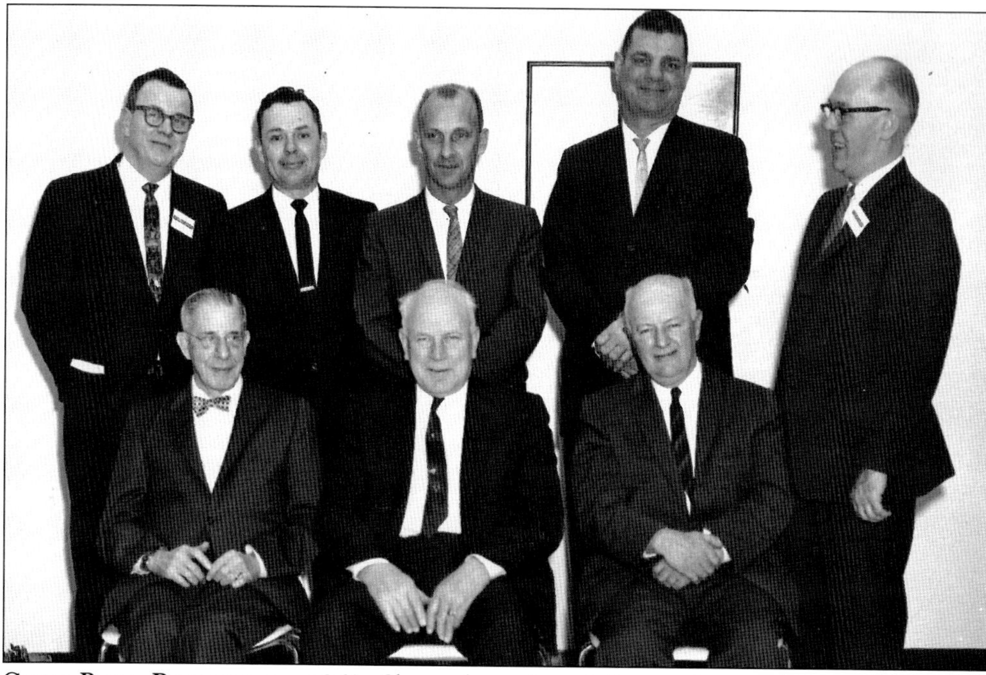

**CHIEF ROHL RETIREMENT, 1963.** Shown here, from left to right, are (first row) fire chief Adolph T. Rohl, Capt. James M. Doran, and Capt. Thomas P. Kane; (second row) unidentified, driver John F. Maille, Capt. Charles L. Mosier, Lt. Paul C. Chapman, and banker Norville Bishop. Following his October 1963 retirement, numerous organizations and individuals feted Rohl, including the Ohio Fire Chief's Association, IAFF Local 165, and U.S. president John F. Kennedy. (Photograph Robert E. Barbian and Associates; courtesy AFD Archive.)

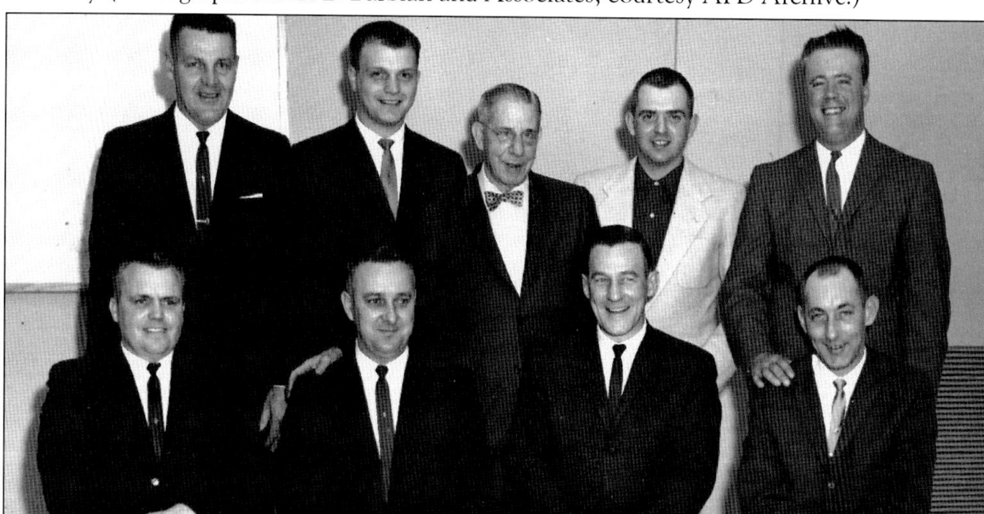

**CHIEF ROHL RETIREMENT, 1963.** Pictured here is retiring 1945–1963 fire chief Rohl (back center) with some of his Ashtabula Fire Department firefighters. The firefighters are, from left to right, (first row) Maynard Meaney, James "Bud" Kirk, Eugene H. Aronhalt, and Donald G. Cuddy; (second row, with Rohl) Donald C. Lloyd, Stanley D. Guerini, Thomas D. Mattern, and Thomas J. Kane, son of veteran firefighter Capt. Thomas P. Kane. (Photograph Robert E. Barbian and Associates; courtesy AFD Archive.)

*Five*

# THE DORAN AND MOSIER ERA

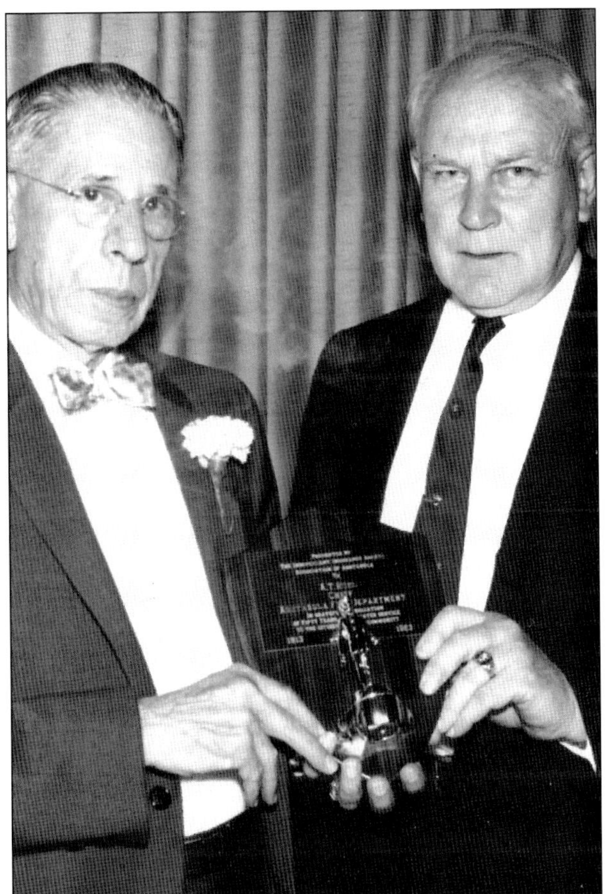

**CHIEF ROHL RETIREMENT PARTY, NOVEMBER 1963.** City manager David P. DeLuca and members of city council were momentarily stunned by a great shock in September 1963, when colorful 18-year fire chief and 50-year department veteran Rohl announced his October 1963 retirement. Here Rohl (left) and successor fire chief James M. Doran, then acting chief, admire Rohl's retirement plaque at a Hotel Ashtabula banquet organized by city employees and sponsored by the Independent Insurance Agents of Ashtabula County. The plaque feted Rohl for his five decades of service to the city between July 1913 and October 1963. James Wentling served as the banquet's main speaker, with longtime Ashtabula insurance agent Francis Herzog serving as master of ceremonies. Noted Rohl of his 50-year career, "I guess I was born to be a fireman. I think I'd do it again." (Photograph Ashtabula Star Beacon; courtesy AFD Archive.)

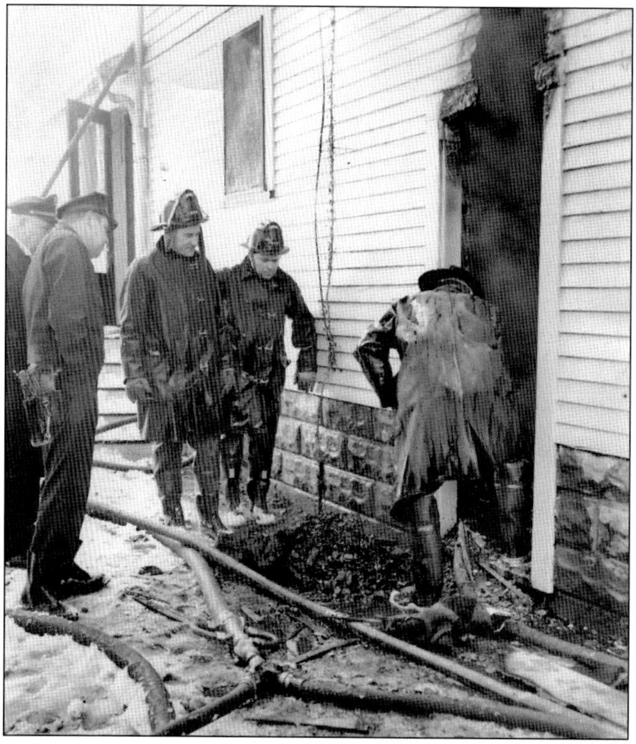

**RESIDENTIAL FIRE, 626 WEST FIFTY-SEVENTH STREET, FEBRUARY 27, 1964.** New city fire chief James M. Doran (extreme left, partially hidden) and city firefighters Eugene H. Aronhalt and Maynard Meaney (center left and right) watch as an unidentified city firefighter (right) sifts through smoldering debris at the gutted David Coy home. The fire, of undetermined origin, severely damaged the house. (Photograph Ashtabula Star Beacon; courtesy AFD Archive.)

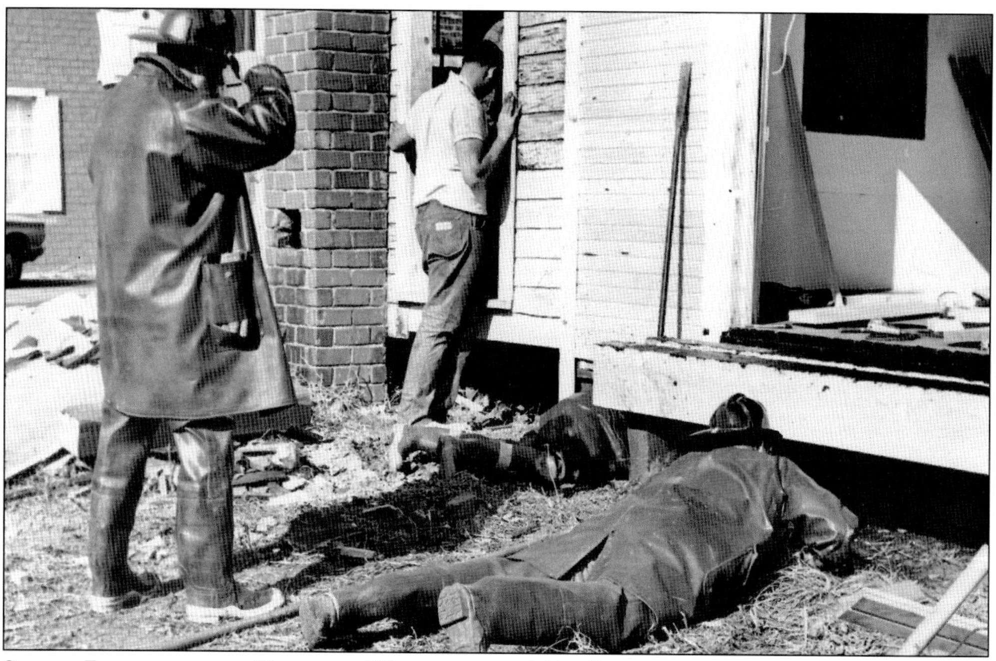

**SEARS, ROEBUCK AND COMPANY WAREHOUSE, 4624 STATE ROAD, SEPTEMBER 25, 1964.** Firefighter Maynard Meaney (left) watches as Capt. Charles L. Mosier (center) and firefighter Stanley D. Guerini (right) scramble with one-inch hoses to reach a small, smoldering rubbish fire in a tight crawl space under the Sears, Roebuck and Company warehouse on State Road. (Photograph Ashtabula Star Beacon; courtesy AFD Archive.)

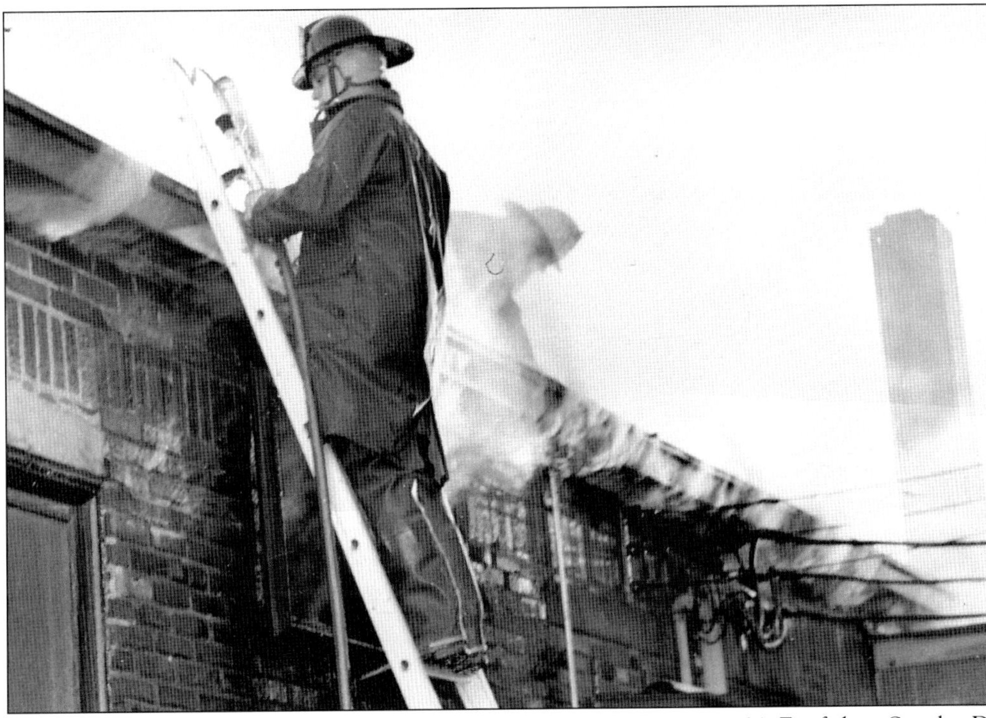

**SWALLOWS RESTAURANT, 1107 PROSPECT AVENUE, NOVEMBER 2, 1964.** Firefighter Stanley D. Guerini prepares to engage a smoky fire at Ashtabula's popular Swallows Restaurant (1939–1966). Burning steaks in an oven broiler touched off the one-and-a-half-hour kitchen blaze, which spread into the adjacent kitchen walls and ceiling. Swallows manager Floyd Redmond praised Ashtabula firefighters for their care in keeping water damage to a minimum. (Photograph Ashtabula Star Beacon; courtesy AFD Archive.)

**NEW SEAGRAVE PUMPER TRUCK.** The booming post–World War II decades was an era of relative prosperity for Ashtabula, which added new fire apparatus in the 1950s and 1960s on the strength of its manufacturing economy and bustling bulk cargo port. Here Capt. Charles L. Mosier showcases the department's new $30,000 c. 1964 Seagrave pumper. A $63,000 100-foot Seagrave aerial ladder truck was added in 1966. (Courtesy Ashtabula Star Beacon.)

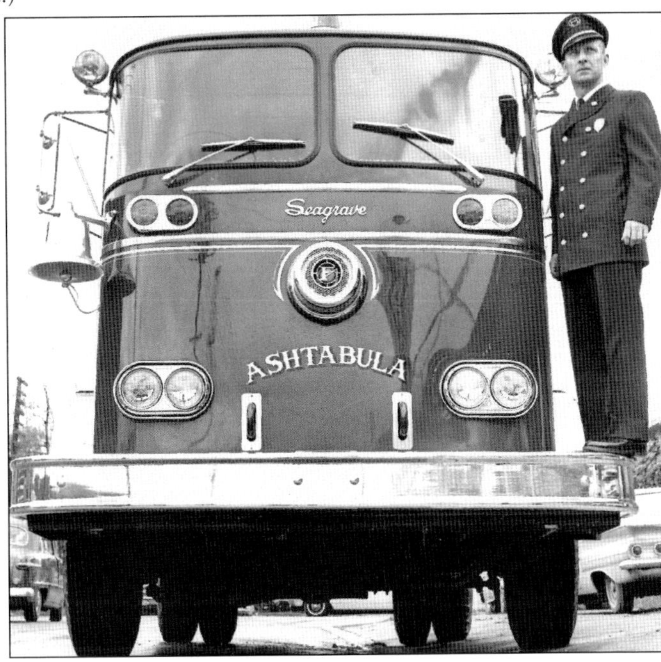

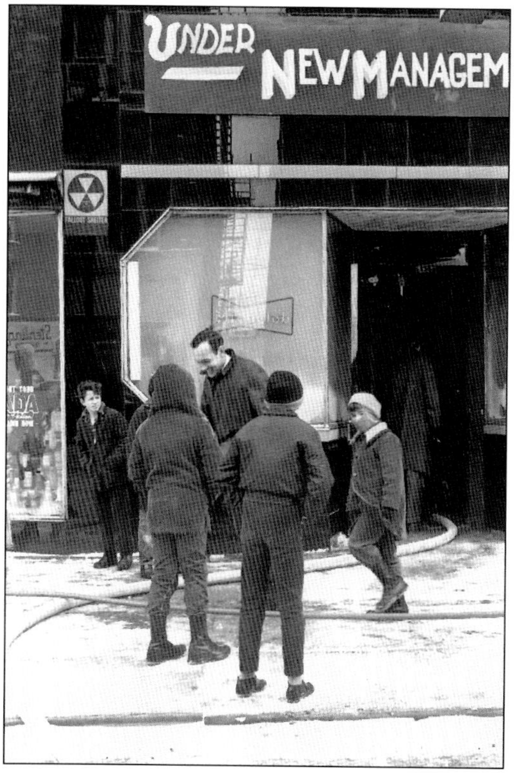


**PENNY FURNITURE FIRE, FEBRUARY 26, 1965.** In the early stages of the $1 million 10-alarm inferno that claimed Main Avenue's Morrison and Nettleton Blocks, there is little hint of the carnage ahead as a single hose line snakes into the Satellite Bar and curious young boys casually mill around on the sidewalk. Thick smoke hid the growing fire in the Nettleton basement. (Photograph Ashtabula Star Beacon; courtesy AFD Archive.)

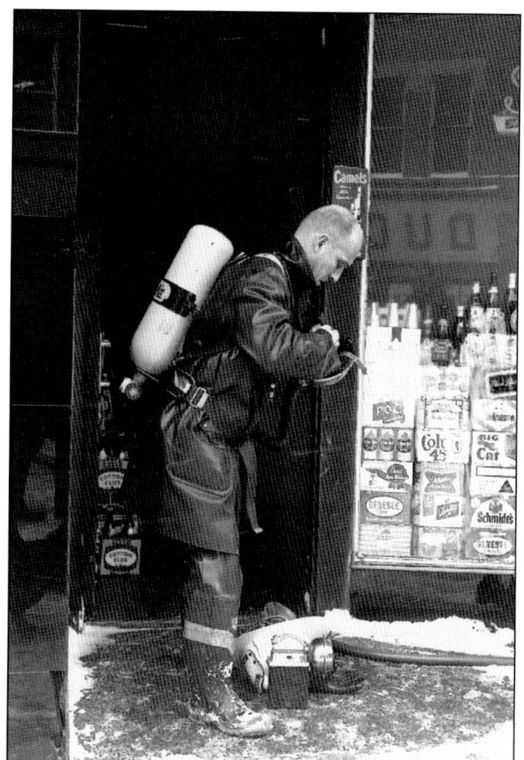


**PENNY FURNITURE FIRE, FEBRUARY 26, 1965.** The first firefighter in, Capt. Charles L. Mosier prepares to enter downtown's Nettleton Block, where intense heat and thick acrid smoke created conditions "impossible to live in." The eight-hour blaze, sparked by a furnace malfunction, was fought by 200 firefighters from Ashtabula City, Ashtabula Township, Saybrook, Jefferson, Kingsville, North Kingsville, Austinburg, Sheffield, Conneaut, and Painesville. (Photograph Ashtabula Star Beacon; courtesy AFD Archive.)

**PENNY FURNITURE FIRE, FEBRUARY 26, 1965.** High winds, near-zero temperatures, and the city's antiquated 34-year-old Seagrave aerial ladder truck made for difficult firefighting during this stubborn $1 million mid-day blaze, which raged out of control for two hours. The fire threatened the neighboring six-story Hotel Ashtabula, prompting its evacuation. The enduring c. 1911 hotel building is slated for mixed-use commercial and residential redevelopment. (Photograph Ashtabula Star Beacon; courtesy AFD Archive.)

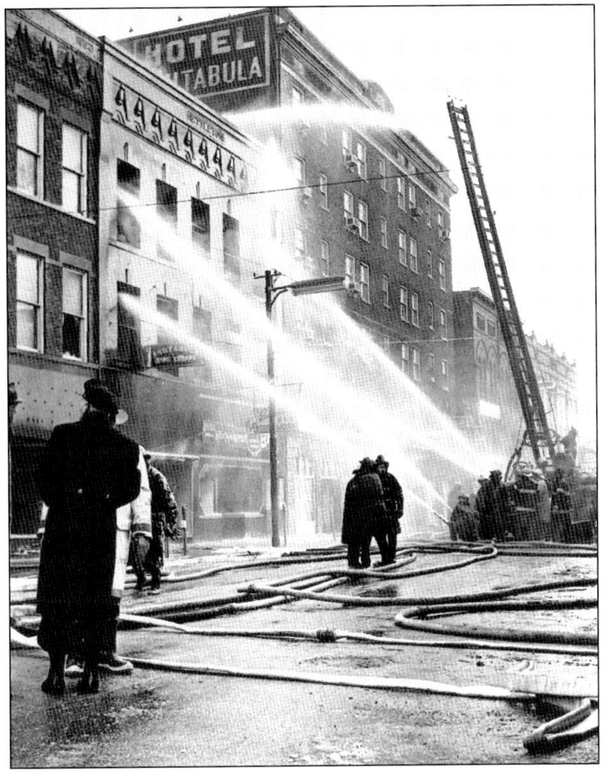


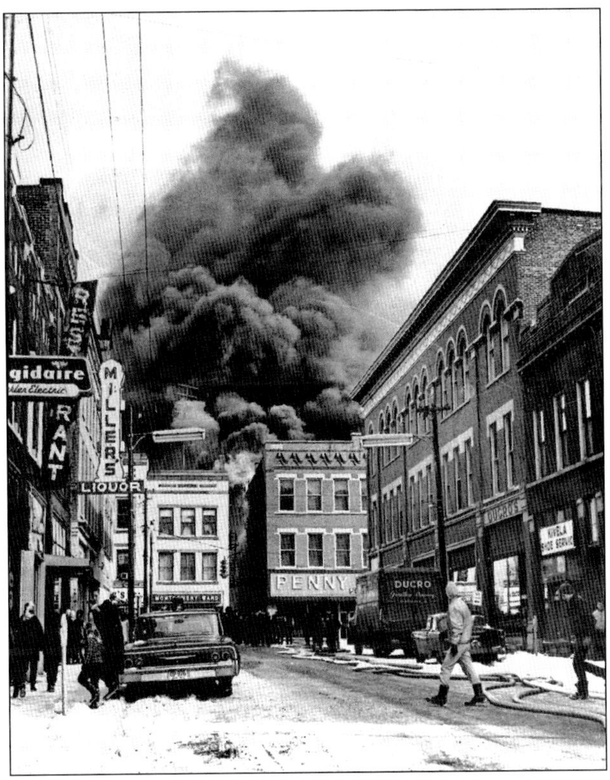


**PENNY FURNITURE FIRE, FEBRUARY 26, 1965.** Looking east on Center Street, thick black smoke pours from the Morrison Block, home to downtown anchor retailer Penny Furniture. Billed as "Ashtabula's worst fire," flames at times rose higher than the neighboring six-story Hotel Ashtabula. Saved by a "wall of water" sprayed on its north wall, the smoke and water-damaged Hotel Ashtabula was closed for two months. (Photograph Ashtabula Star Beacon; courtesy AFD Archive.)

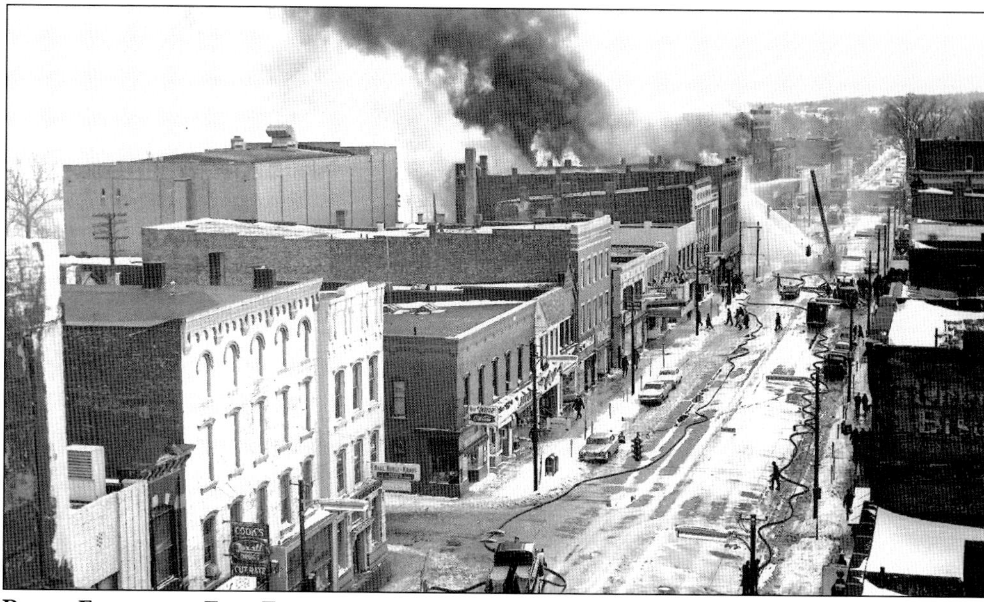

**PENNY FURNITURE FIRE, FEBRUARY 26, 1965.** Looking southeast from the five-story Carlisle-Allen department store, smoke and flames fanned by a strong west wind billow over downtown Ashtabula. Nineteen fire companies representing 10 departments battled this tenacious $1 million blaze in the Morrison and Nettleton Blocks, which displaced several downtown businesses including Penny Furniture Company, the Satellite Bar, Ashtabula Wine Shoppe, and Ashtabula Business College. (Photograph Ashtabula *Star Beacon*; courtesy AFD Archive.)

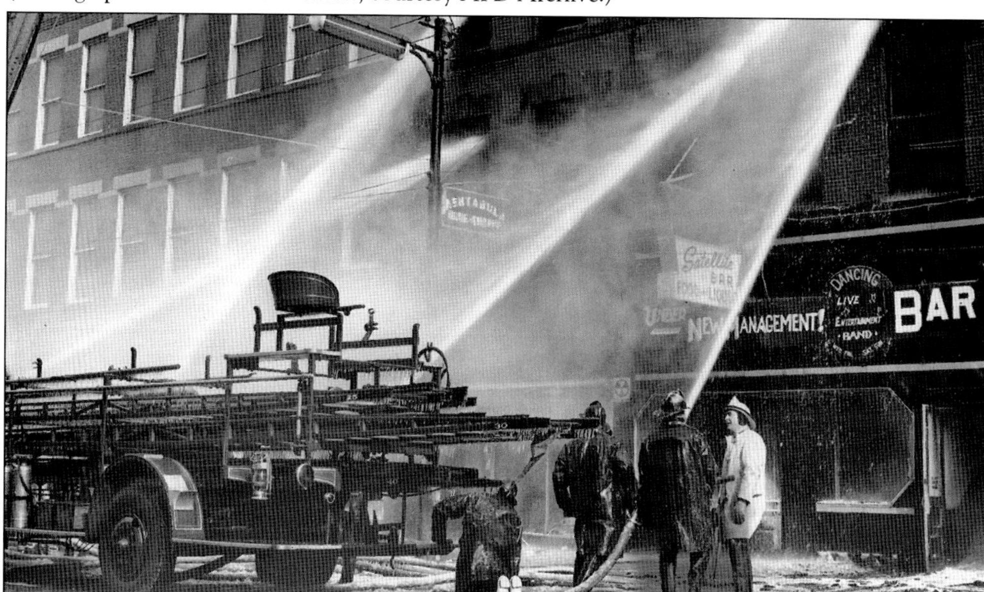


**PENNY FURNITURE FIRE, FEBRUARY 26, 1965.** An out-of-town fire chief talks with Ashtabula firefighters manning hose lines near the city's c. 1930 Seagrave aerial ladder truck. The performance of Painesville's modern aerial ladder truck hastened replacement of Ashtabula's aging and outdated 75-foot Seagrave. In October 1966, the Ashtabula Fire Department took delivery on a new $63,000 Seagrave aerial ladder truck. (Photograph Ashtabula *Star Beacon*; courtesy Ashtabula County District Library.)

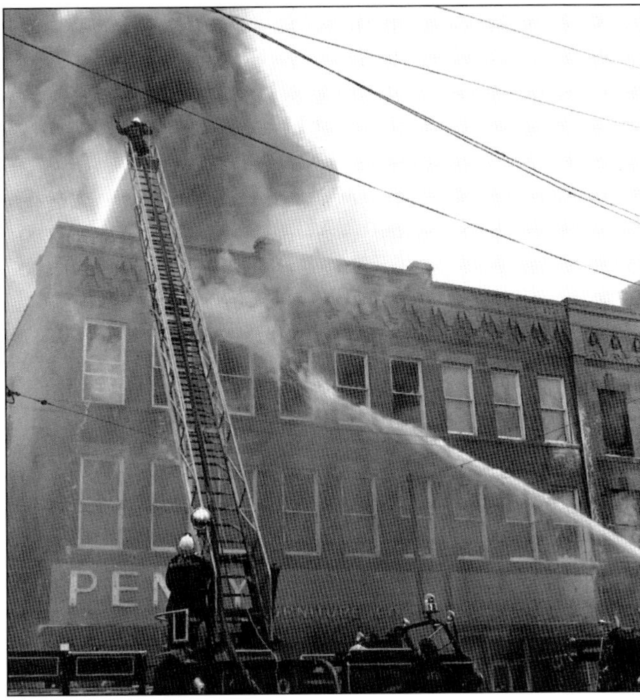


**PENNY FURNITURE FIRE, FEBRUARY 26, 1965.** The performance of Painesville's modern aerial ladder truck, seen here dousing downtown's fiery Morrison Block, underscored Ashtabula's need to replace its c. 1930 Seagrave aerial hook and ladder truck. In October 1966, the department took delivery on a state-of-the-art 100-foot Seagrave aerial truck—just weeks before a four-building inferno occurred across Main Avenue from this fire scene. (Photograph Ashtabula Star Beacon; courtesy Ashtabula County District Library.)

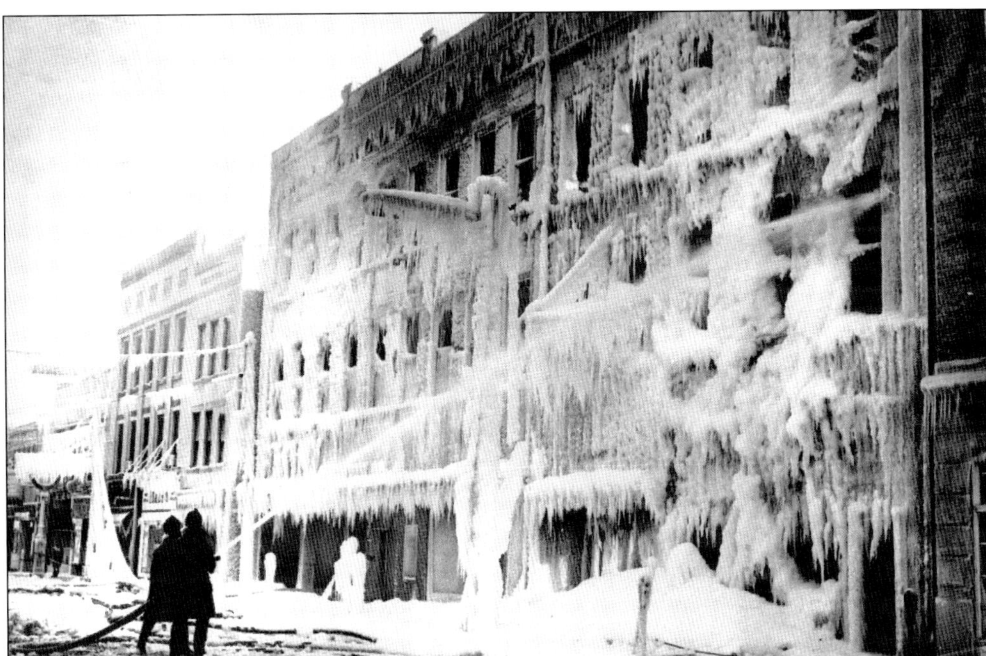

**PENNY FURNITURE FIRE, FEBRUARY 27, 1965.** Temperatures near zero and two million gallons of water created a thick layer of ice on the charred remains of downtown Ashtabula's Morrison and Nettleton Blocks. An estimated 10,000 spectators visited Main Avenue over several days to photograph the fire and its icy aftermath. The *Star Beacon* increased print runs to meet "high reader interest and sales demand." (Photograph Ashtabula Star Beacon; courtesy AFD Archive.)

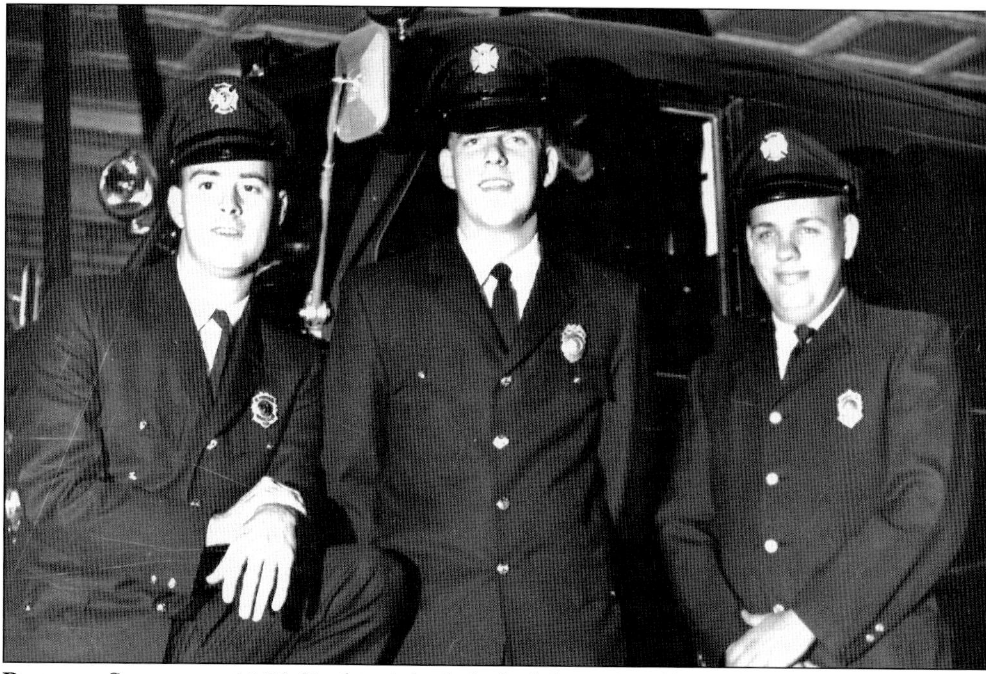

**ROOKIES, SEPTEMBER 1965.** Rookie Ashtabula firefighters Gerald C. Thurston (left), Charles P. (Chuck) Brockway (center), and Richard K. Good are seen in Uptown's truck bay. Brockway served 33 years with the department, promoted to engineer in 1971, lieutenant in 1974, and captain in 1976. Between 1973 and 1998, Brockway logged 25 years with the department's Fire Inspection Bureau, spearheading fire prevention and arson investigation activities. (Photograph Ashtabula Star Beacon; courtesy Charles P. Brockway.)

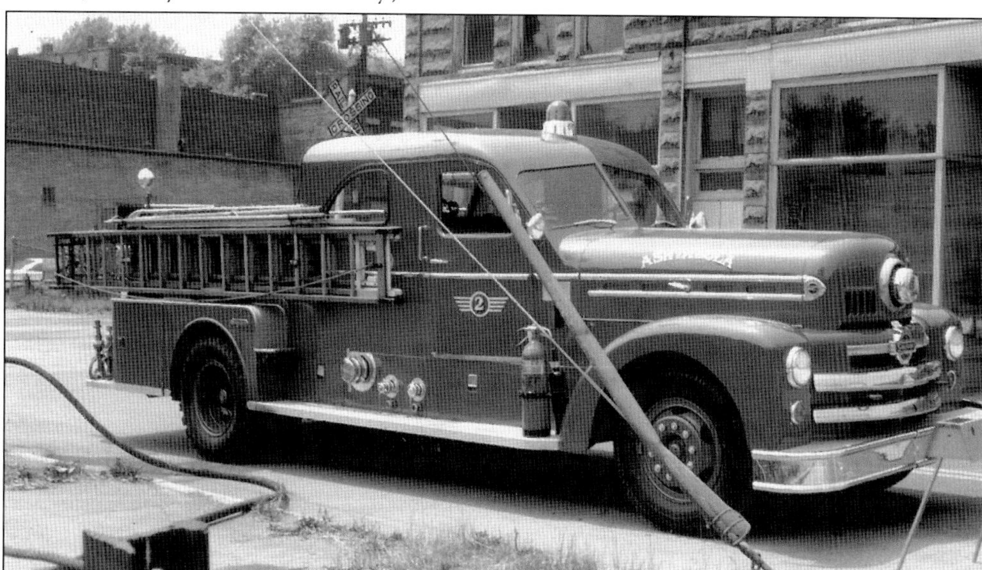

**ENGINE 2, JUNE 1966.** This 1956 Seagrave pumper was stationed at the Harbor Fire Station on West Ninth Street. In the days of the old two-station department, odd numbers were assigned to the headquarters station in downtown Ashtabula, with even numbers assigned to the Harbor Fire Station. (Courtesy AFD Archive.)

**BROWN FURNITURE FIRE—OCTOBER 29, 1966.** Downtown's Main Avenue retailing district was struck by its second devastating fire in 20 months when high winds spread this $1 million early morning blaze, destroying the Giantos, Sanborn, D. L. Davis, and Warren Hotel buildings. Looking south from near Center Street, flames from the four-story Warren Hotel threaten the c. 1904 Ducro Block, which survived the multi-building blaze. (Photograph Ashtabula Star Beacon; courtesy Charles P. Brockway.)

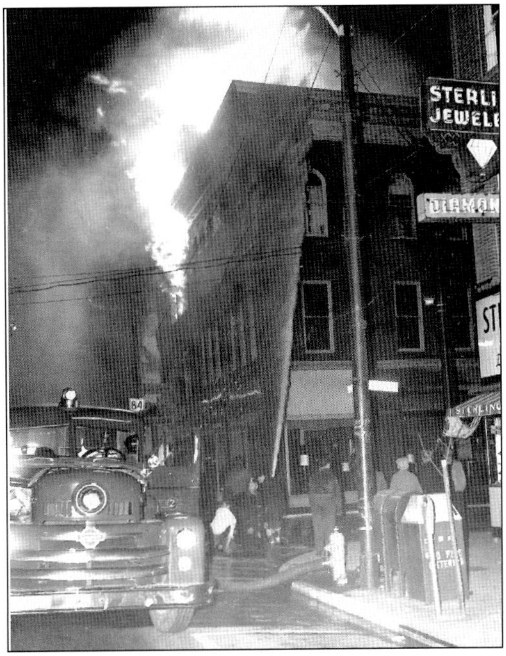

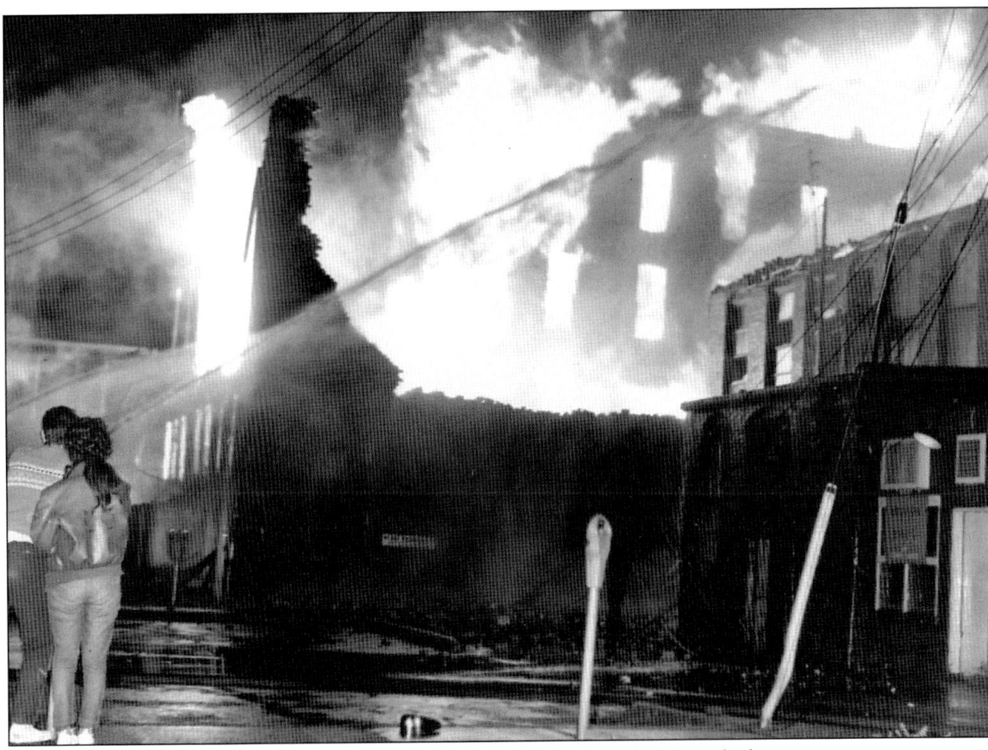

**BROWN FURNITURE FIRE—OCTOBER 29, 1966.** As bystanders watch, hose streams are sprayed on the collapsed ruins of the three-story D. L. Davis Block (center) and the fiery blaze consuming the neighboring four-story Warren Hotel. Visible for miles, the fire showered downtown with dangerous wind-driven embers. Early Ashtabula firemen on the scene were assisted by civilians who joined in the fight. (Photograph Ashtabula Star Beacon; courtesy Charles P. Brockway.)

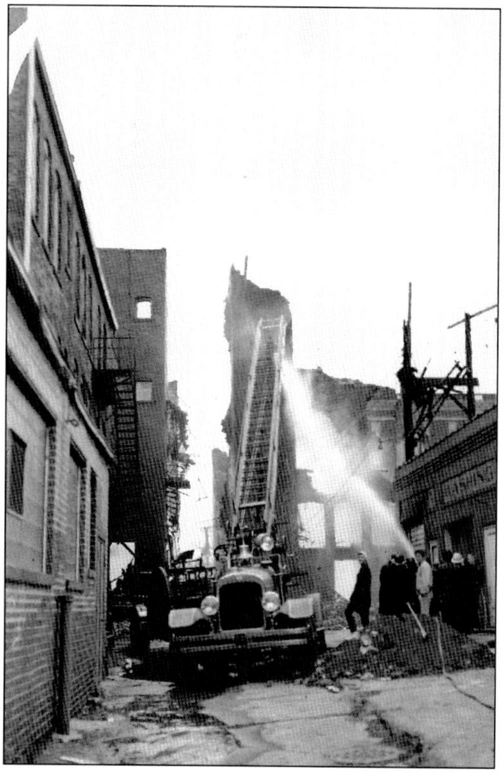

**BROWN FURNITURE FIRE—OCTOBER 29, 1966.** Although brought under control within two hours, fire crews took no chances at hot spot flare-ups as they poured 12 lines of water into the smoldering debris of the Warren Hotel and D. L. Davis Block. The fire's ferocity saw Ashtabula's recently retired c. 1930 Seagrave ladder truck pressed into service one final time. (Photograph Ashtabula Star Beacon; courtesy Charles P. Brockway.)

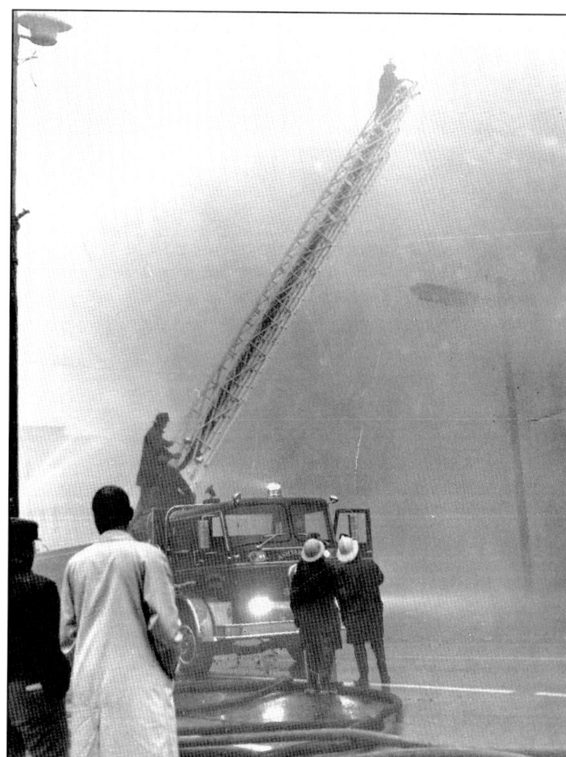

**BROWN FURNITURE FIRE—OCTOBER 29, 1966.** The smoldering remains of the D. L. Davis Block and the Warren Hotel cast a smoky daybreak pall over Main Avenue as Ashtabula firefighters use their new 100-foot Seagrave aerial truck to douse hot spots. Reported the *Star Beacon*, "firemen were credited with doing an outstanding job of fighting one of the worst fires in city history." (Photograph Ashtabula Star Beacon; courtesy Charles P. Brockway.)

**BROWN FURNITURE FIRE—OCTOBER 29, 1966.** Already doomed by the planned urban renewal demolition of the block bounded by Main Avenue, Center Street, Park Avenue and West Forty-eighth Street, fire hastened the demise of the D. L. Davis Block (left) and the Warren Hotel (center). Ducro Furniture (1853–1973) in the Ducro Block relocated to new Park Avenue quarters in 1970. (Photograph Ashtabula Star Beacon; courtesy Charles P. Brockway.)

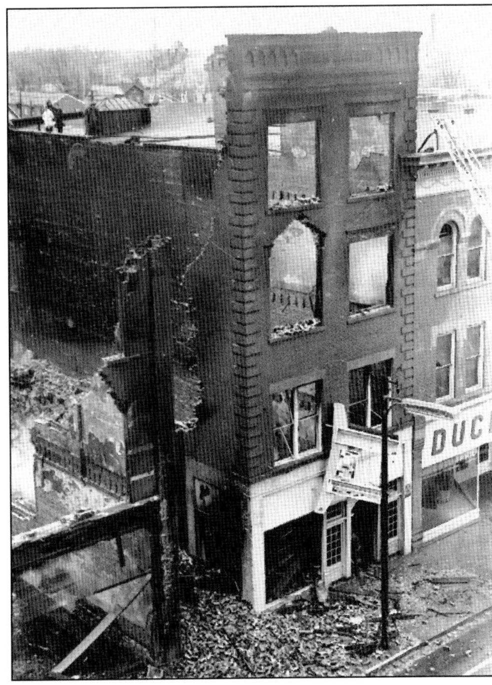

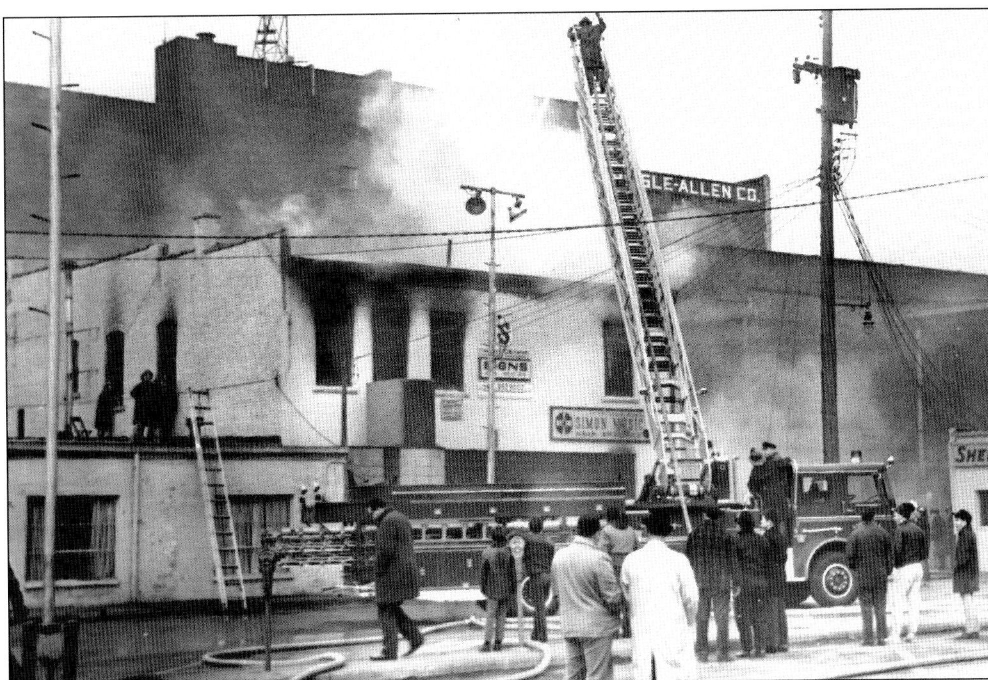

**SIMON MUSIC STORE, 228 PROGRESS PLACE—FEBRUARY 20, 1967.** This two hour, late afternoon arson blaze extensively damaged the Simon Music Store and the upstairs United Steelworkers union hall. With downtown Ashtabula ravaged by massive multi-building fires in 1965 and 1966, firefighters feared the blaze would spread to the Carlisle-Allen department store and J. J. Newberry variety store. Ashtabula City and Township firefighters contained the blaze. (Courtesy Ashtabula Star Beacon.)

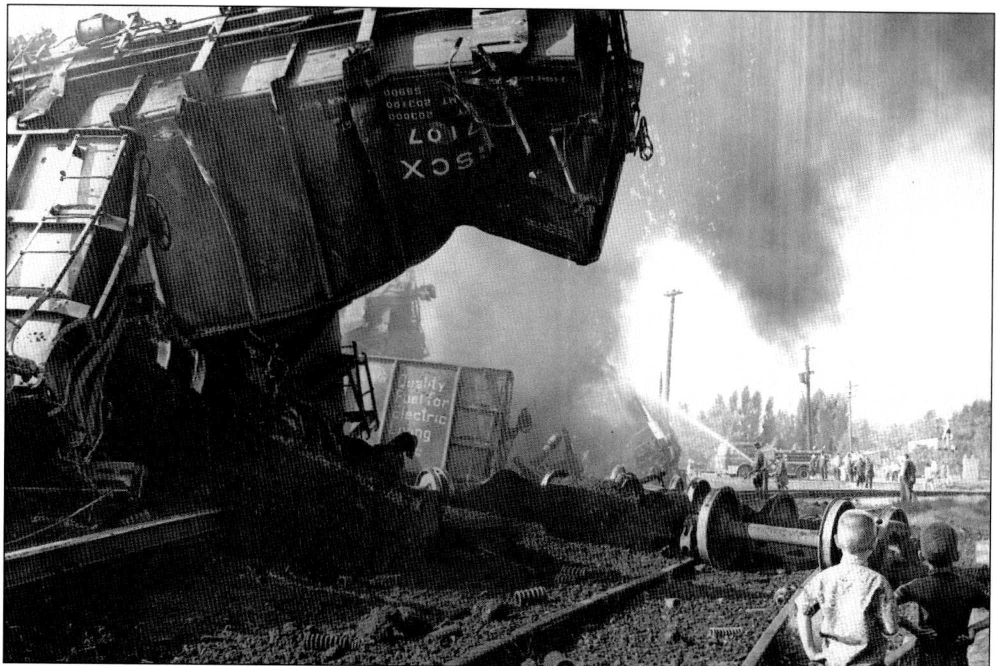

**PENN CENTRAL DERAILMENT—SEPTEMBER 24, 1968.** Losing brake power shortly after leaving the Carson Yard in Plymouth Township, a runaway Penn Central coal train moving at nearly 65 miles per hour derailed in this massive multimillion-dollar crash near West Thirty-second Street. Five diesel engines and 66 of the train's 93 fully loaded coal cars derailed. Uptown and Harbor firefighters and mutual aid crews from Ashtabula and Saybrook Townships spent four hours dousing fires sparked by the derailed diesel engines. Hundreds of curious area residents turned out to view the spectacular derailment. A massive clean-up effort, dubbed "Operation Jackstraw," was mounted by the Penn Central to clear the accident scene, including 60,000 tons of coal dumped along a 1,000-foot stretch of track. (Courtesy Ashtabula Star Beacon.)

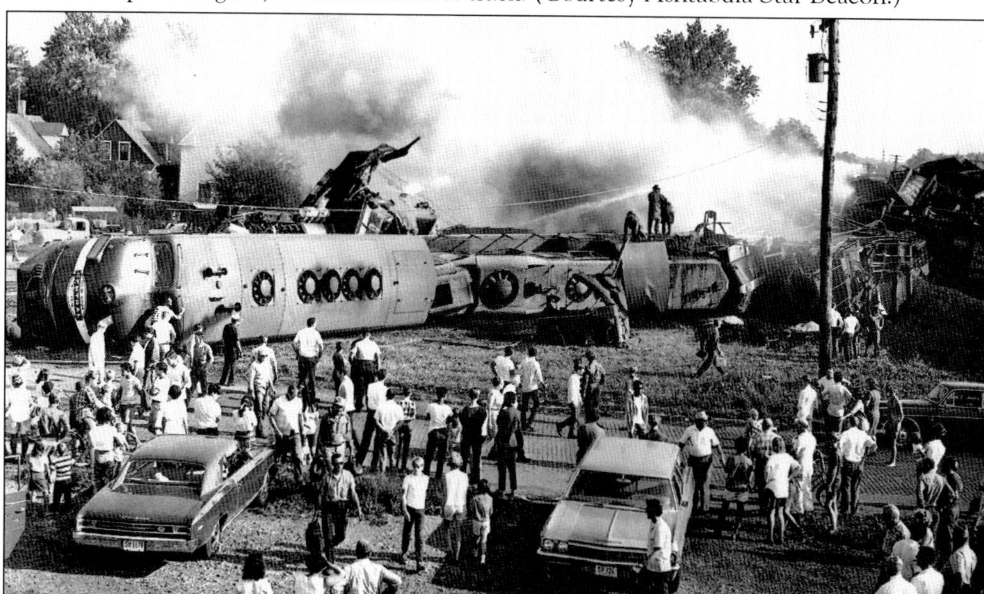

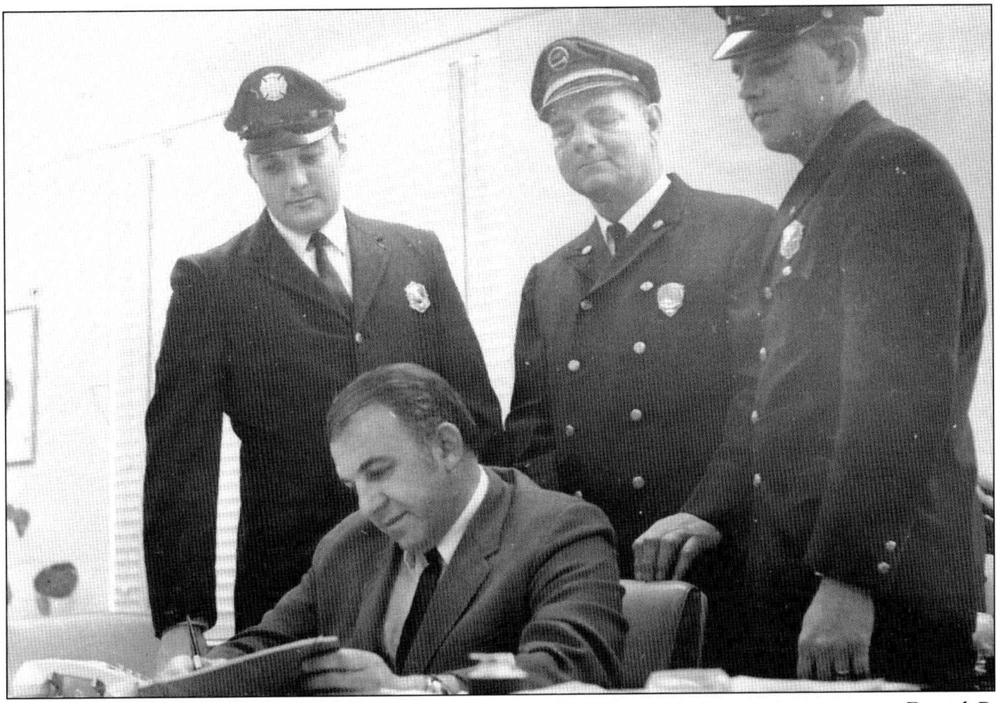

**FIRST IAFF CONTRACT, 1969.** Following six months of negotiations, city manager David P. DeLuca signs Ashtabula's first contract with IAFF Local 165. Founded in 1919, Local 165 had finally gained recognition as the sole bargaining representative for Ashtabula firefighters regarding wages, hours, and working conditions. Local 165 officials included president Frank Acierno, negotiating committee chairman Paul C. Chapman, and secretary-treasurer Charles P. Brockway. (Photograph Ashtabula Star Beacon; courtesy Charles P. Brockway.)

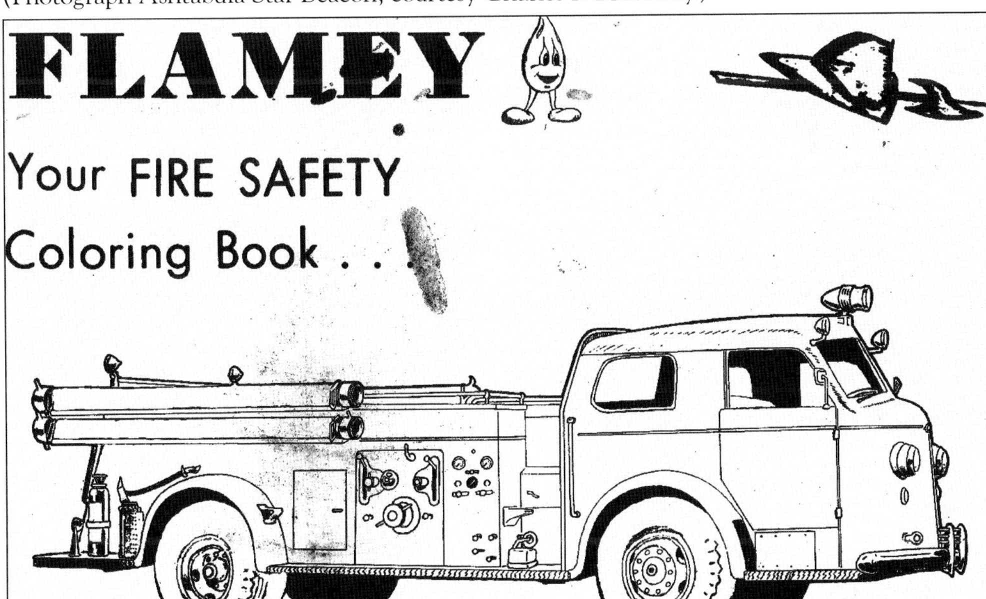


**FLAMEY, YOUR FIRE SAFETY COLORING BOOK.** This complimentary Fire Prevention Bureau coloring book was an early foray into children's fire safety programs. (Courtesy AFD Archive.)

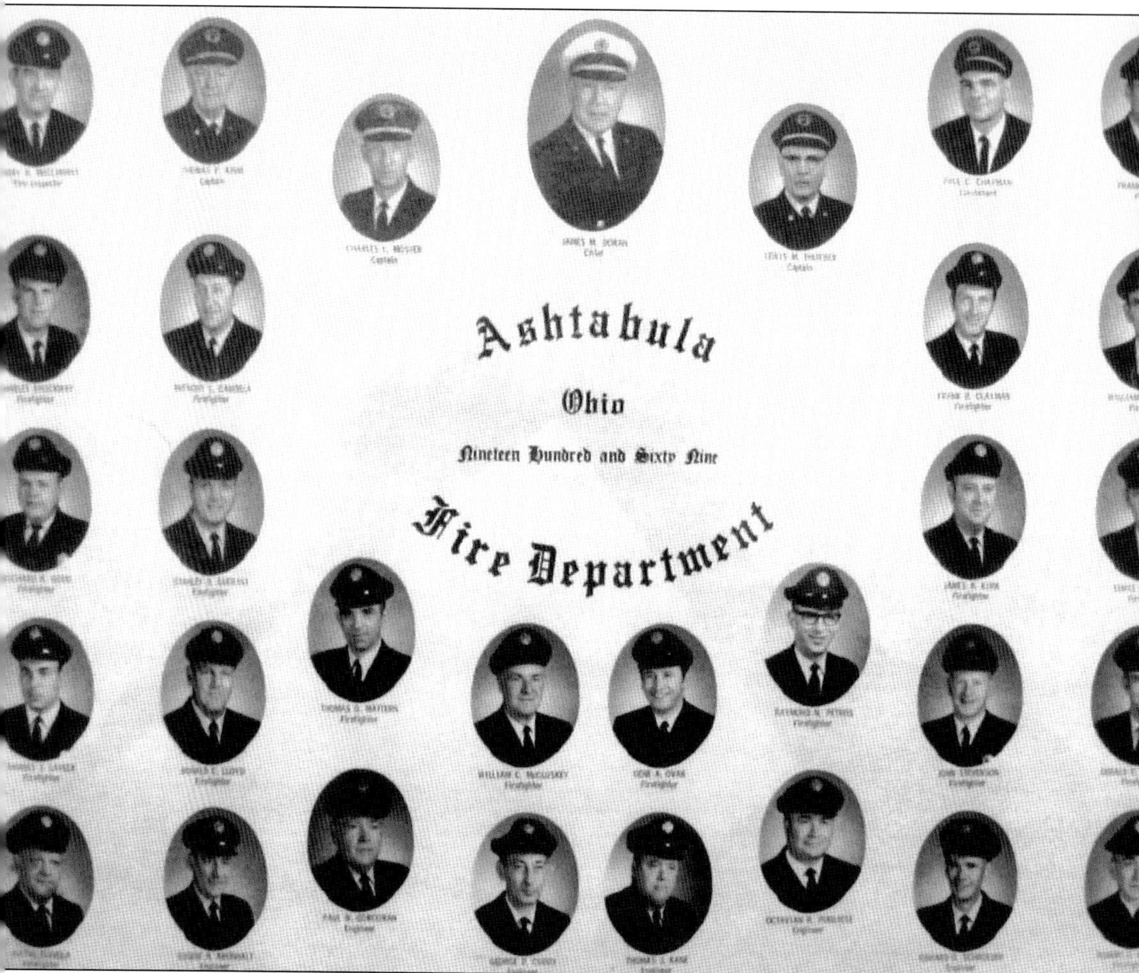


**ASHTABULA FIRE DEPARTMENT ROSTER, 1969.** Pictured here is the department's 31-man, two-station 1969 roster, as photographed by Detroit-based Allen Studios. They are, from left to right (first row) firefighter Waino Toivola, engineers Eugene H. Aronhalt, Paul W. Corcoran, George D. Cuddy, Thomas J. Kane, Octavian R. Pugliese, and Edward G. Schroeder, and firefighter Robert C. Wilcom; (second row) firefighters Thomas J. Laveck, Donald C. Lloyd, Thomas D. Mattern, William C. McCluskey, Gene A. Ovak, Raymond N. Petros, John Stevenson, and Gerald C. Thurston; (third row) firefighters Richard K. Good, Stanley D. Guerini, James A. Kirk, and Lewis W. Laszlo; (fourth row) firefighters Charles P. Brockway, Anthony L. Candela, Frank R. Clayman, and William J. DeSorbes; (fifth row) fire inspector Lt. Harry A. McClimans, Capt. Thomas P. Kane, Capt. Charles L. Mosier, fire chief James M. Doran, Capt. Lewis Thurber, Lt. Paul C. Chapman, and firefighter Frank Acierno. (Courtesy AFD Archive.)

**FIRE PREVENTION WEEK, 1971.** Taking a proactive approach to firefighting, the Ashtabula Fire Department established its Fire Prevention Bureau in 1967. Bureau inspector between 1967 and 1973, 23-year veteran Lt. Harold A. (Harry) McClimans and "junior fire marshal" Sparky promote a Fire Prevention Week initiative—old coffee cans filled with bicarbonate of soda for fighting kitchen fires. McClimans later served as Ashtabula Township's first full-time fire chief. (Courtesy Ashtabula Star Beacon.)

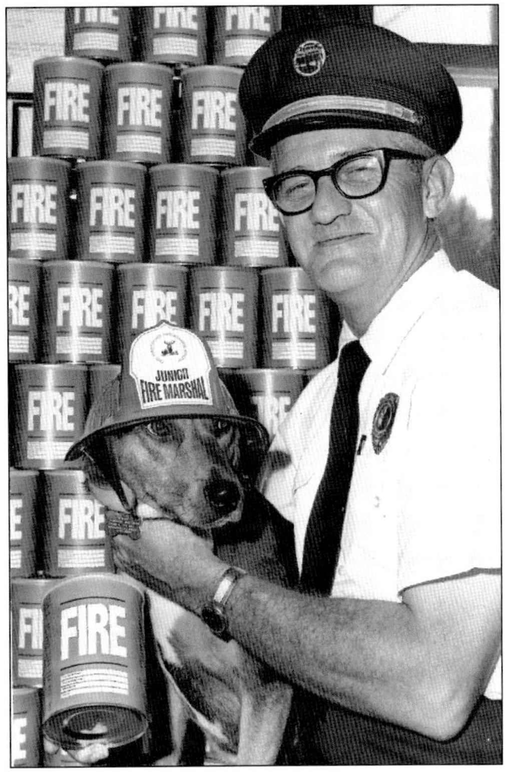


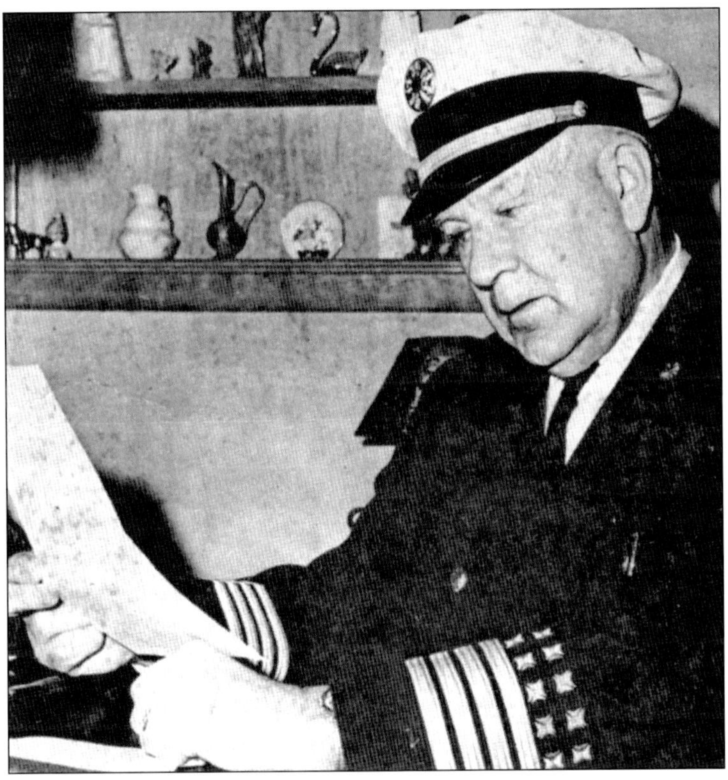

**CHIEF DORAN RETIREMENT, 1971.** Fire chief James M. Doran looks over paperwork at his home as he prepares to retire after a nearly 54-year career with the Ashtabula Fire Department between 1917 and 1971, including a seven-year tenure as fire chief. Doran was instrumental in the January 1919 organization of Ashtabula's IAFF Local 165, one of the first Ohio firefighter locals to receive state sanction. (Photograph Ashtabula Star Beacon; courtesy AFD Archive.)

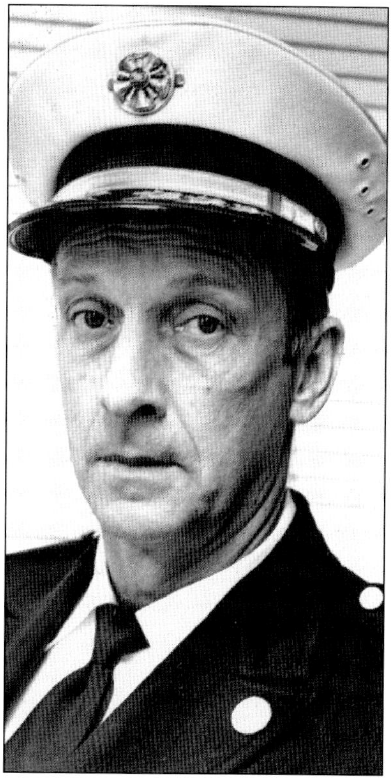

**Charles L. Mosier Appointed Fire Chief—June 1971.** Following James Doran's retirement, city manager George E. Ducro III appointed Charles L. Mosier as Ashtabula's new fire chief. A 40-year department veteran between 1947 and 1987, Mosier logged 16 years as fire chief. Mosier's major accomplishments included the $600,000 development of Station One in 1974–1975 and the purchase of four new pumpers in 1971, 1976, and 1986. (Courtesy Ashtabula Star Beacon.)

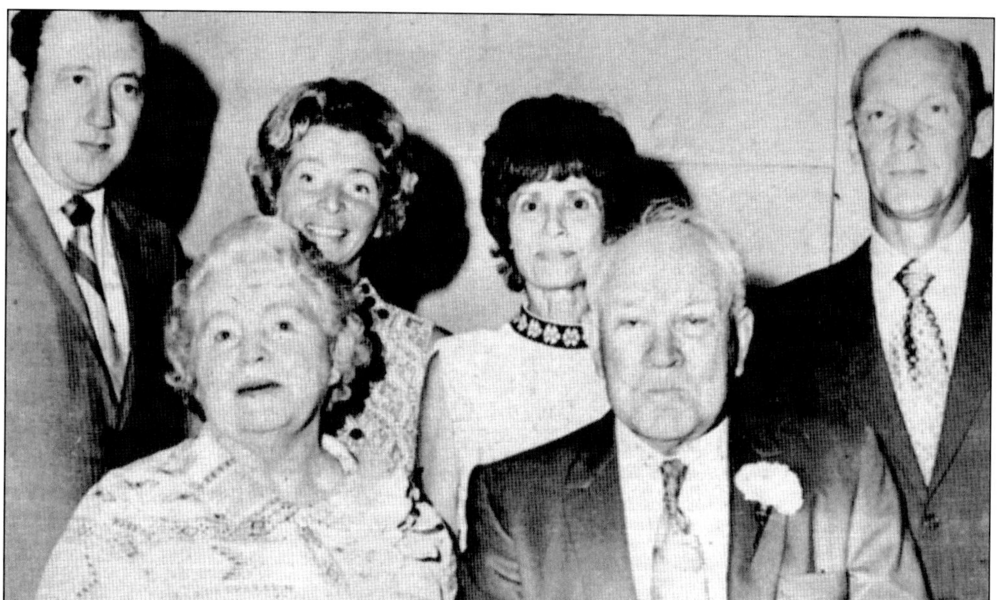

**Chief Doran Appreciation Dinner, June 1971.** The Hotel Ashtabula ballroom was the venue for an appreciation dinner in honor of retired fire chief and 54-year department veteran James M. Doran (front right), seated with his wife Vera. Among those attending (from left) were city manager George E. Ducro III and his wife Pauline, Celeste Mosier, and successor fire chief Charles L. Mosier. (Photograph Ashtabula Star Beacon; courtesy AFD Archive.)

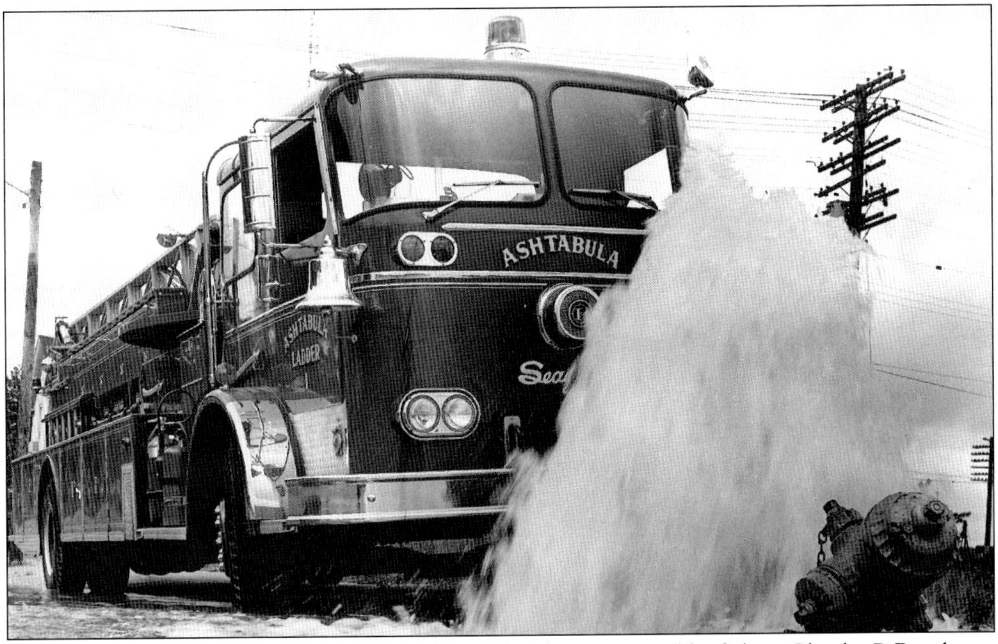


"I KNEW IT WAS THERE," JUNE 1972. Such was the observation of firefighter Charles P. Brockway (1965–1998) when he spawned this bubbling gusher after bumping a Park Avenue hydrant while parking the department's *c.* 1966 Seagrave ladder truck across from the headquarters Uptown Fire Station. In addition to the broken hydrant, the accident caused minor fender damage. (Photograph Ashtabula Star Beacon; courtesy Charles P. Brockway.)

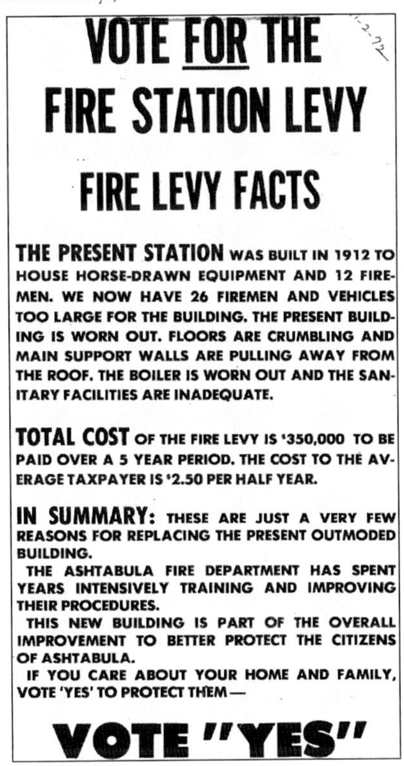


"VOTE *For* THE FIRE STATION LEVY," NOVEMBER 1972. In 1969 and 1970, voters defeated 20-year bond referendums to build a state-of-the-art headquarters station to replace the department's *c.* 1912 Uptown Fire Station. Hoping to persuade city voters to endorse a .988-mill levy, the Independent Insurance Agents of Ashtabula published this Ashtabula *Star Beacon* advertisement. Voters sent the five-year $350,000 bond issue to a 5,378 to 4,156 defeat. (Courtesy AFD Archive.)

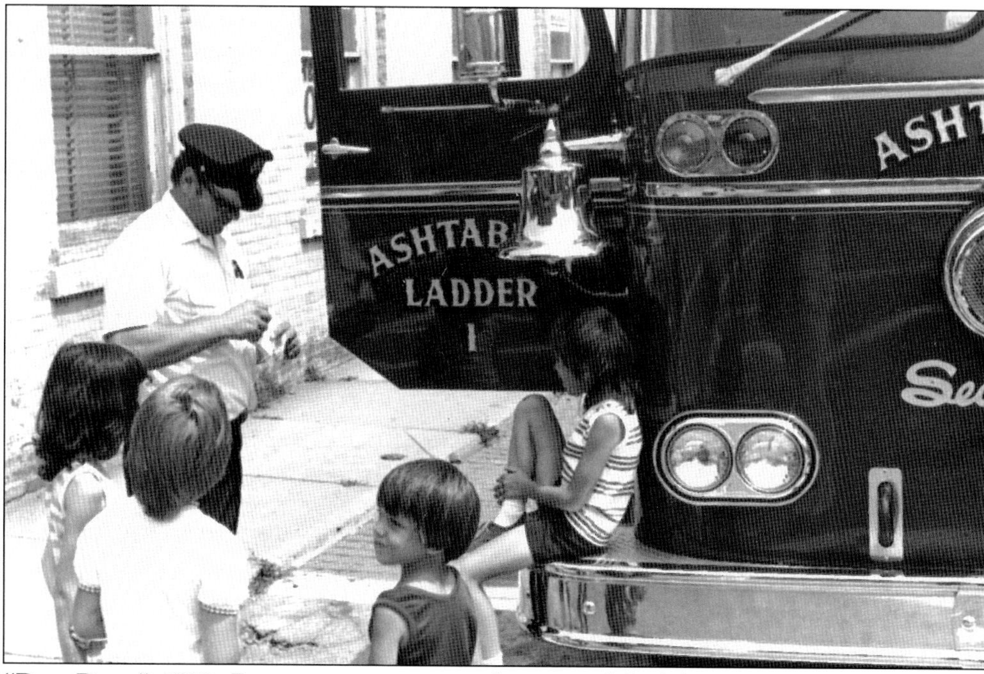


**"Dog Days," 1973.** Eager city youngsters gather around firefighter Steve Deak (1971–1996) outside downtown's Uptown Fire Station during Dog Days, a summer Fire Prevention Bureau community relations promotion that included the distribution of fire prevention brochures and mingling with the public, especially children. Deak, certified as an EMT instructor in 1975, prepares to bandage a scraped knee for the young girl sitting on Ladder One's bumper. (Courtesy AFD Archive.)

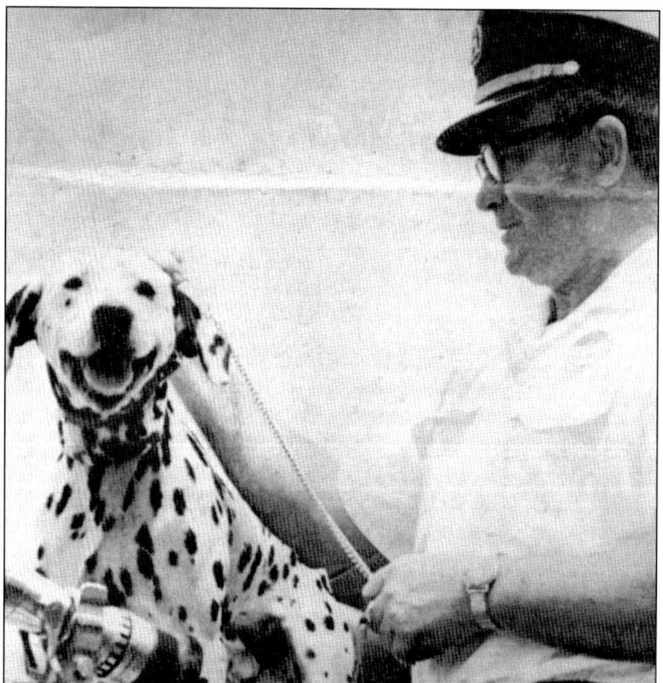

**Royal Spukel the First, July 1973.** Capt. Octavian "Barney" Pugliese pets new Uptown fire dog Royal Spukel the First, nicknamed "Spook," shortly after the four-year-old purebred Dalmatian joined the Ashtabula Fire Department. Donated to the department, Spook was soon farmed out to a rural family, having "acquired a number of habits over his life that are not compatible with the fire service." (Photograph Ashtabula Star Beacon; courtesy AFD Archive.)

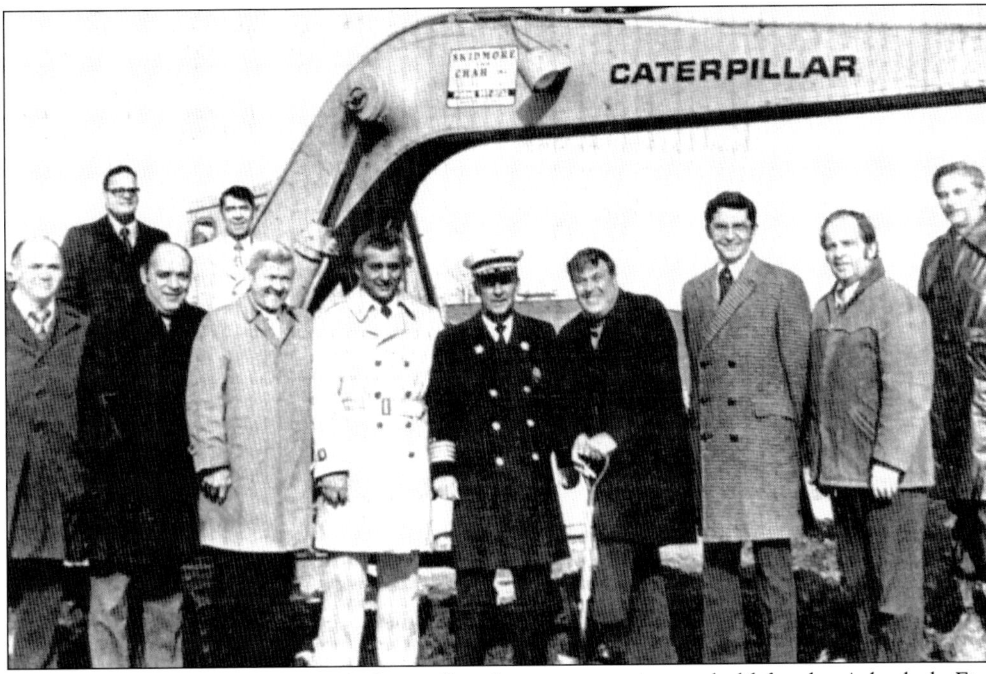


**GROUNDBREAKING, APRIL 1974.** Groundbreaking ceremonies are held for the Ashtabula Fire Department's new $536,776 headquarters Station One facility, 4326 Main Avenue. Attending (first row, from left) were city auditor Robert Sabo, councilman Dominic Colucci, council vice president Clifford McClure, councilman Peter DiAngelo, fire chief Charles L. Mosier, city manager Edward F. Bento, council president Joseph DelPriore, councilman Richard Peura, and Ashtabula architect Peter Chismar. Standing in back are councilmen Harold Gran and Charles Barrett. (Photograph Ashtabula Star Beacon; courtesy AFD Archive.)

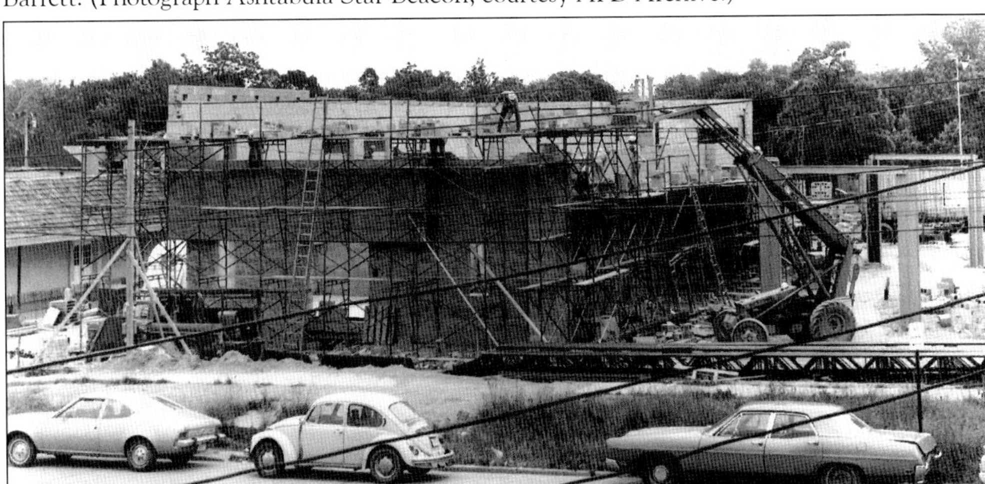

**STATION ONE CONSTRUCTION, SUMMER 1974.** Construction gets underway on Station One, 4326 Main Avenue. Ashtabula issued five-year councilmatic notes to finance construction, using federal revenue sharing money to repay the debt. By the mid-1970s, the Ashtabula Fire Department encompassed a 7.4-square-mile service area with nearly 25,000 residents. A roster of 39 Harbor and Uptown firefighters manned four engine companies and one ladder company. (Photograph G. K. Lewis Creative Photography; courtesy AFD Archive.)

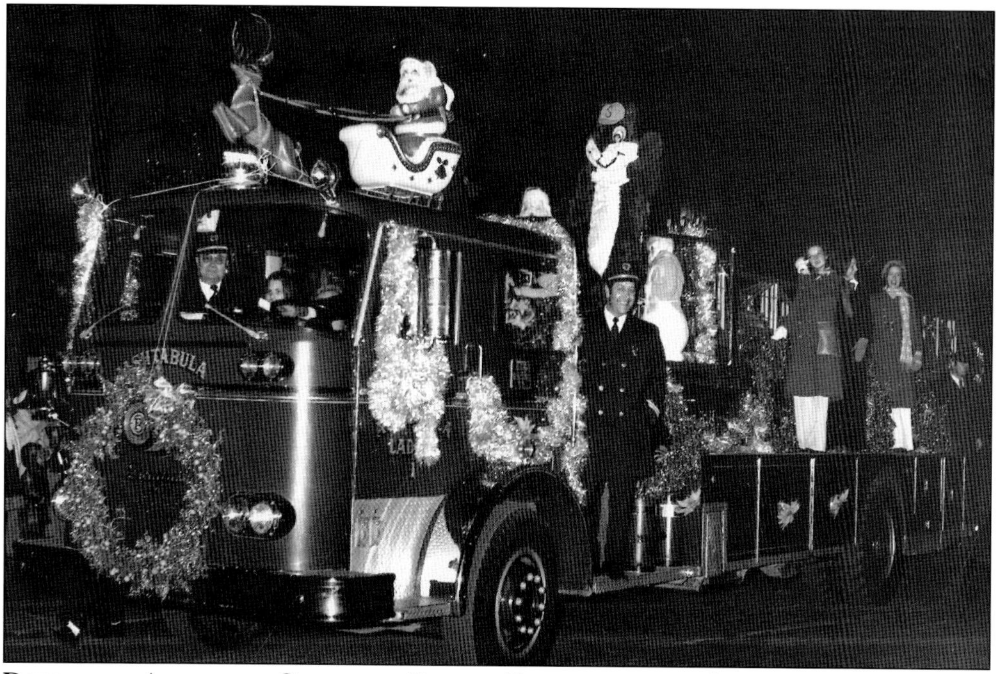

**Downtown Ashtabula Christmas Parade, December 1974.** The Fire Prevention Bureau's ever-smiling Capt. Anthony L. Candela (center) brightens the Ashtabula Fire Department's festively decorated *c.* 1966 Seagrave aerial ladder truck during the annual Christmas parade down Main Avenue, a popular event cosponsored by the Downtown Merchant's Association and the Ashtabula Jaycees. Seen in Ladder One's cab at left is Capt. Octavian R. "Barney" Pugliese (1953–1983). (Photograph Ashtabula Star Beacon; courtesy AFD Archive.)

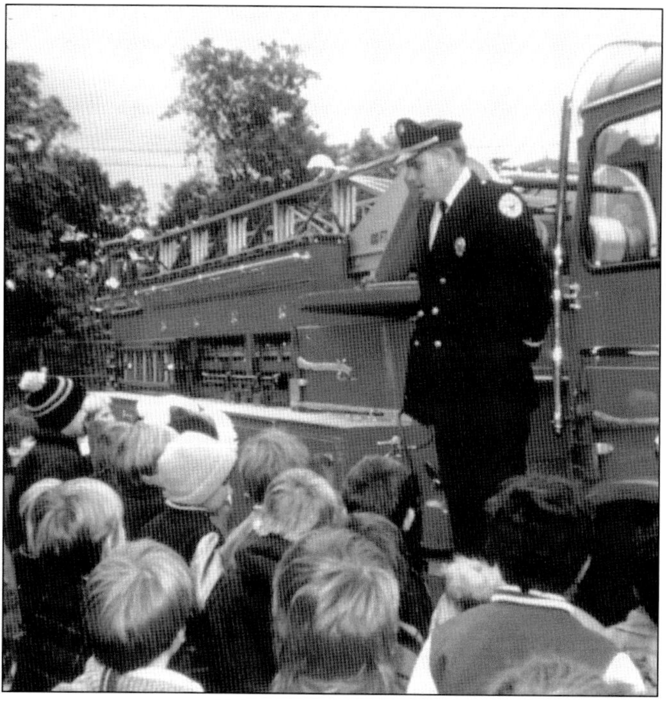

**Fire Prevention Bureau Program, October 1974.** Standing on the Ashtabula Fire Department's *c.* 1966 Seagrave aerial ladder truck, Lt. Charles P. Brockway makes a Fire Prevention Bureau presentation to Roman Catholic elementary students at St. Joseph School, 3400 Lake Avenue. Wrote principal Sister Julie Demchak to fire chief Charles L. Mosier, "Lt. Brockway's . . . presentation . . . was most informative and we thought him a beautiful representative for your department." (Courtesy Charles P. Brockway.)

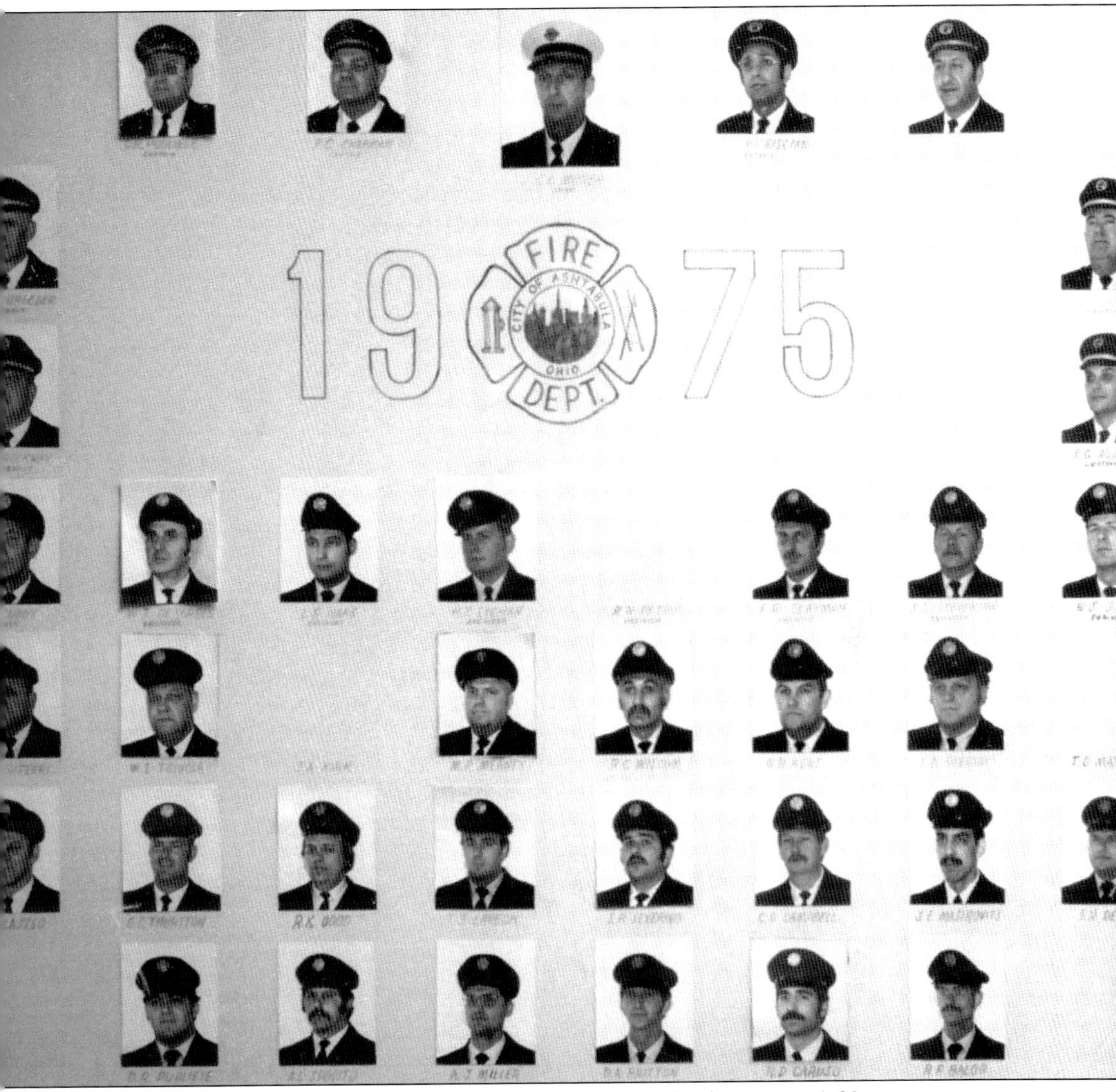

**Ashtabula Fire Department Roster, 1975.** Pictured is the department's 39-man, two-station roster in 1975. A deep local recession, rustbelt decline, and a city fiscal emergency would spur the 1988 closure of the city's Harbor Fire Station and the downsizing of the department's firefighter roster. Shown here are, from left to right, (first row) firefighters D. R. Pugliese, A. L. Sposito, A. J. Miller, D. A. Britton, R. D. Caruso, and R. F. Balog; (second row) firefighters L. W. Laszlo, G. C. Thurston, R. K. Good, T. J. Laveck, S. P. Severino, C. G. Campbell, J. E. Masirovits, and S. N. Deak. (third row) firefighters A. J. Ucciferri, W. I. Toivola, M. P. Meaney, R. C. Wilcom, G. D. Kent, and S. D. Guerini; (fourth row) Engineers G. D. Cuddy, W. J. DeSorbes, L. G. Hare, H. T. Leehan, F. R. Clayman, J. S. Stevenson, and N. J. Jepson; (fifth row) Lts. C. P. Brockway and F. G. Acierno; (sixth row) Lts. E. G. Schroeder and T. J. Kane; (seventh row) Capt. O. R. Pugliese, Capt. Paul C. Chapman, fire chief Charles L. Mosier, Capt. G. C. Matrisciano, and Capt. A. L. Candela. Not pictured are engineer R. N. Petros and firefighters J. A. Kirk and T. D. Mattern. (Courtesy AFD Archive.)

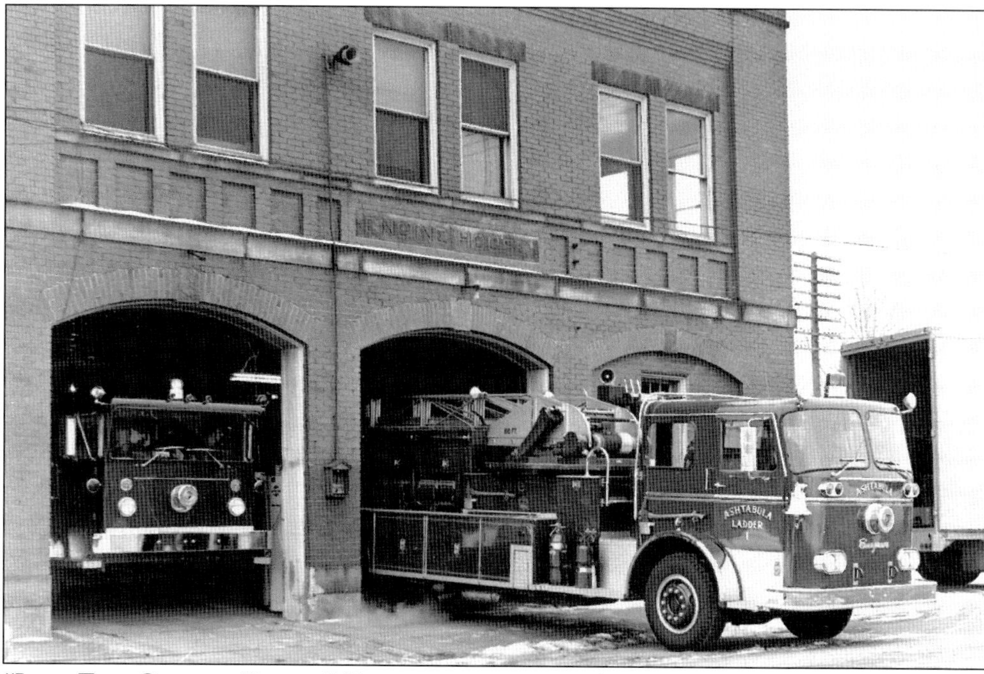

**"Last Time Out the Doors," February 13, 1975.** Urban renewal and the need for a larger, state-of-the-art facility spurred the closure of Ashtabula's *c.* 1912 headquarters Uptown Fire Station, 4758 Park Avenue. Ashtabula's *c.* 1966 Seagrave ladder leads the apparatus procession to the city's new Station One headquarters, 4326 Main Avenue. Uptown was demolished in March–April 1975, in part to widen West Forty-eighth Street. (Photograph Ashtabula Star Beacon; courtesy AFD Archive.)

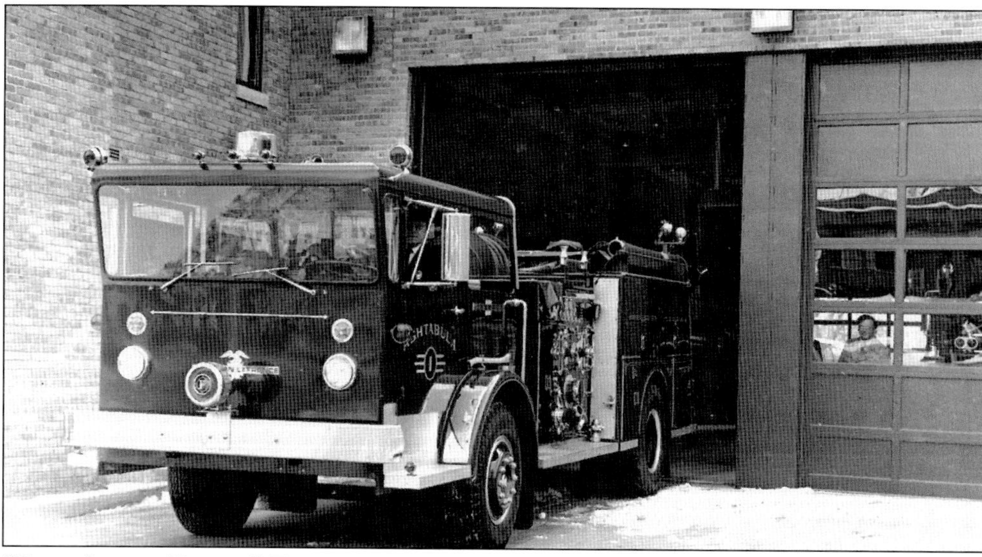

**"First In the Doors," February 13, 1975.** The department's *c.* 1971 American-LaFrance pumper backs into the truck bay at the new $536,776 Station One headquarters on Main Avenue. Noted fire chief Charles Mosier of the new era, "The day . . . of the fire hall as a lounge area is past in today's modern fire operation concept. The old card-playing, red-suspendered fireman is from a different era." (Photograph Ashtabula Star Beacon; courtesy AFD Archive.)

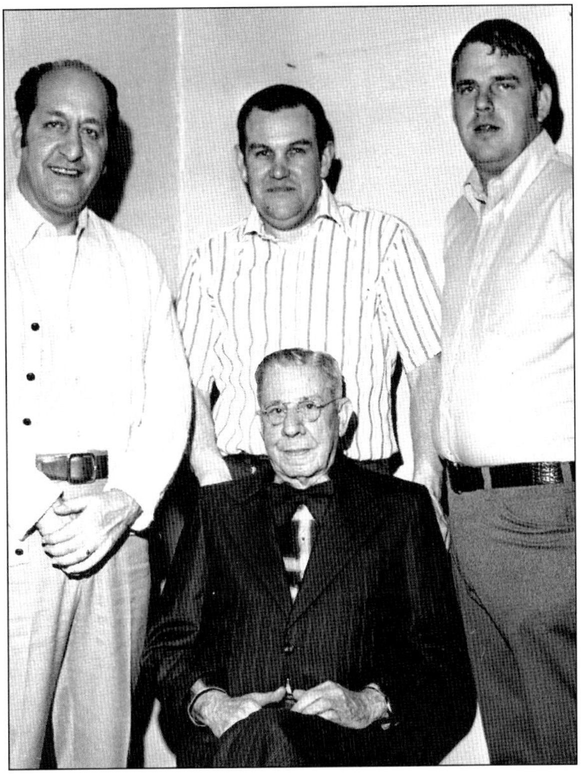

**UPTOWN FIRE STATION FAREWELL PARTY, FEBRUARY 1975.** Active and retired Ashtabula firefighters gathered at the idled *c.* 1912 Uptown Fire Station to share memories and bid the venerable brick firehouse farewell. The gathering, organized by firefighter Gary D. Kent (center rear), included retired fire chief Adolph T. Rohl (seated) and Fire Prevention Bureau inspectors Capt. Anthony L. Candela (left) and Lt. Charles P. Brockway (right). (Courtesy AFD Archive.)

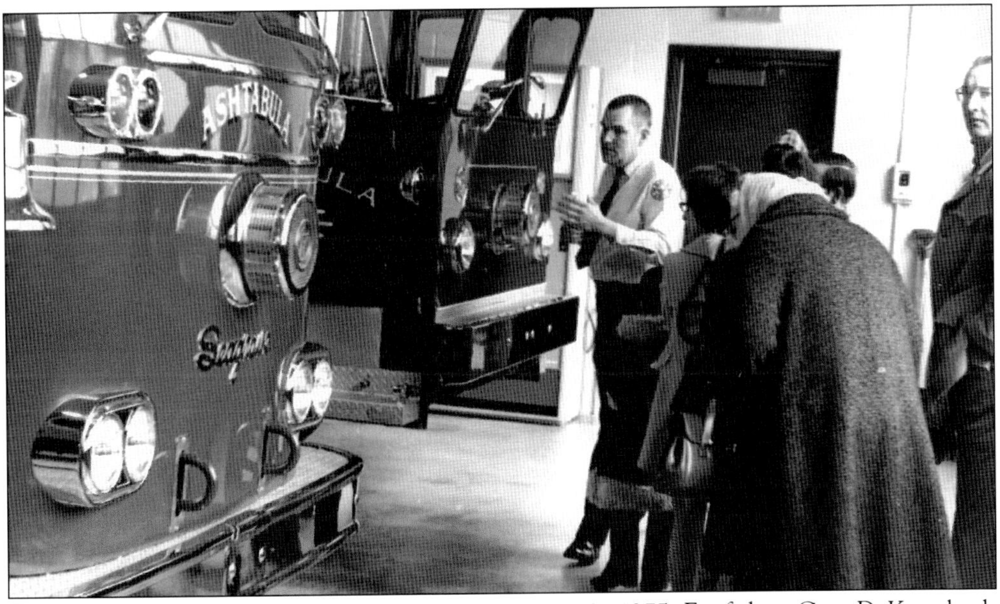


**STATION ONE COMMUNITY OPEN HOUSE, MARCH 1–2, 1975.** Firefighter Gary D. Kent leads a tour through Station One's expansive truck bay during a gala weekend open house, which included firefighter-guided tours and refreshments served by the Ashtabula Exchange Club and other local service organizations. A downtown urban renewal development, Station One's construction was made possible by federal revenue sharing funds. (Photograph G. K. Lewis Creative Photography; courtesy AFD Archive.)

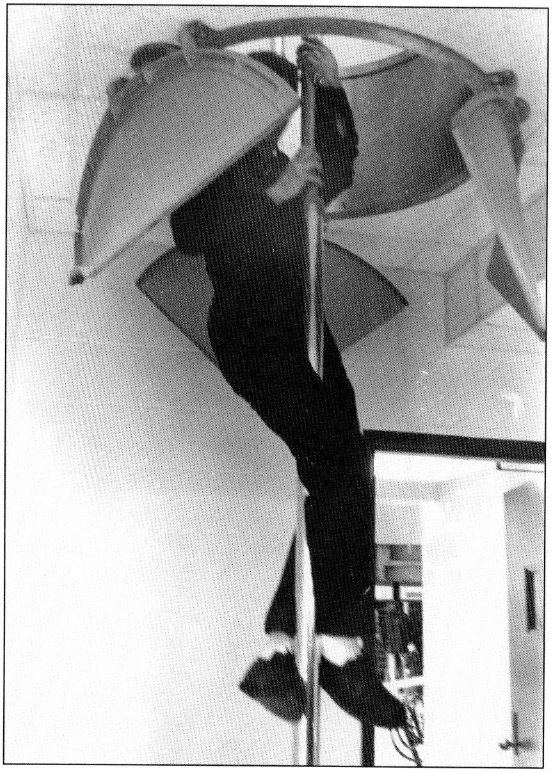

**STATION ONE COMMUNITY OPEN HOUSE, MARCH 1–2, 1975.** A city firefighter demonstrates the art of fire pole sliding during a community open house at the Ashtabula Fire Department's new $536,776 headquarters facility. Downward pressure on the fire pole by a firefighter on the second floor opens the three hinged safety doors that close the floor hole when the pole is not in use. (Photograph G. K. Lewis Creative Photography; courtesy AFD Archive.)

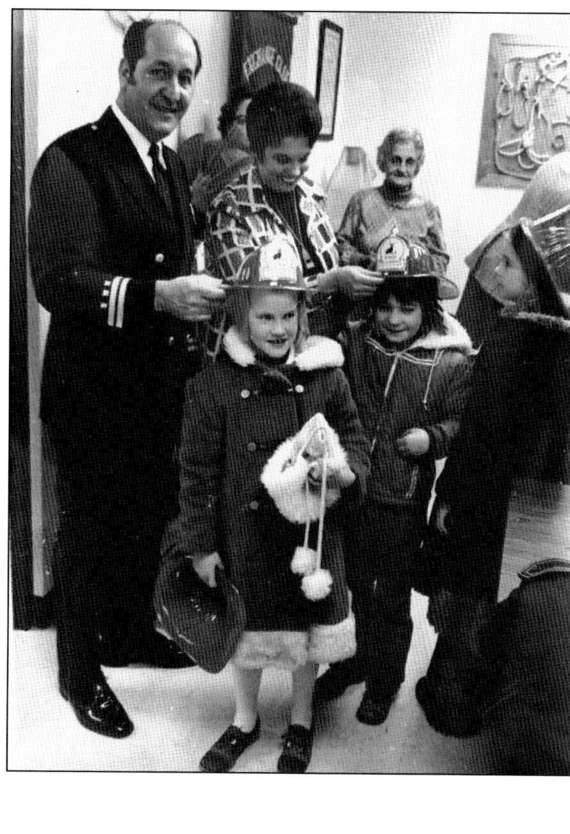

**STATION ONE COMMUNITY OPEN HOUSE, MARCH 1-2, 1975.** Lt. Anthony L. Candela and his wife Madeline hand out red plastic fire helmets to visiting children during a two-day community open house at Station One. Candela, an 18-year department veteran praised as a "leader" and "man of dignity," tragically died of a heart attack in February 1976 at age 43. (Photograph G. K. Lewis Creative Photography; courtesy Charles P. Brockway.)

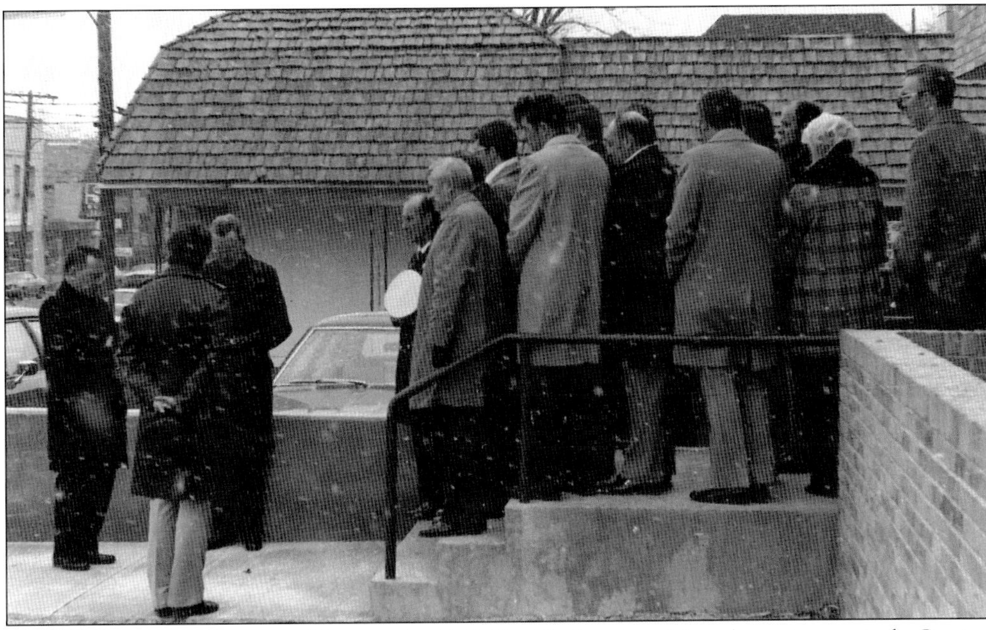

**STATION ONE DEDICATION, MARCH 1, 1975.** Huddled against bitter late winter winds, Station One dedication attendees included fire chief Charles Mosier, city council members, city manager Edward Bento, and Ashtabula architect Peter G. Chismar. Prayers were offered by Rev. Bernard Vacca of Our Lady of Mount Carmel Catholic Church and Rev. George Johnson of First Presbyterian Church. (Photograph G. K. Lewis Creative Photography; courtesy AFD Archive.)

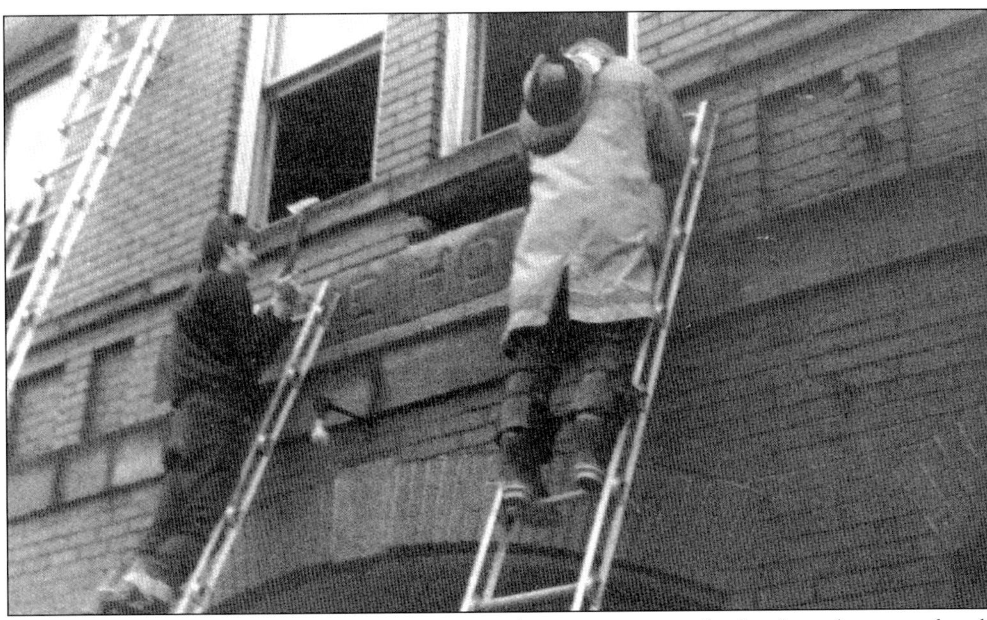

**SAVING A PIECE OF HISTORY, APRIL 1975.** Chipping away at the brick and mortar facade of the Ashtabula Fire Department's c. 1912 Uptown Fire Station on Park Avenue, firefighters James Masirovits (left) and Steve Deak work to remove the station's inset stone "Engine House" marker prior to its demolition. The marker was later installed on a decorative wall outside the department's new headquarters Station One facility on Main Avenue. (Courtesy AFD Archive.)

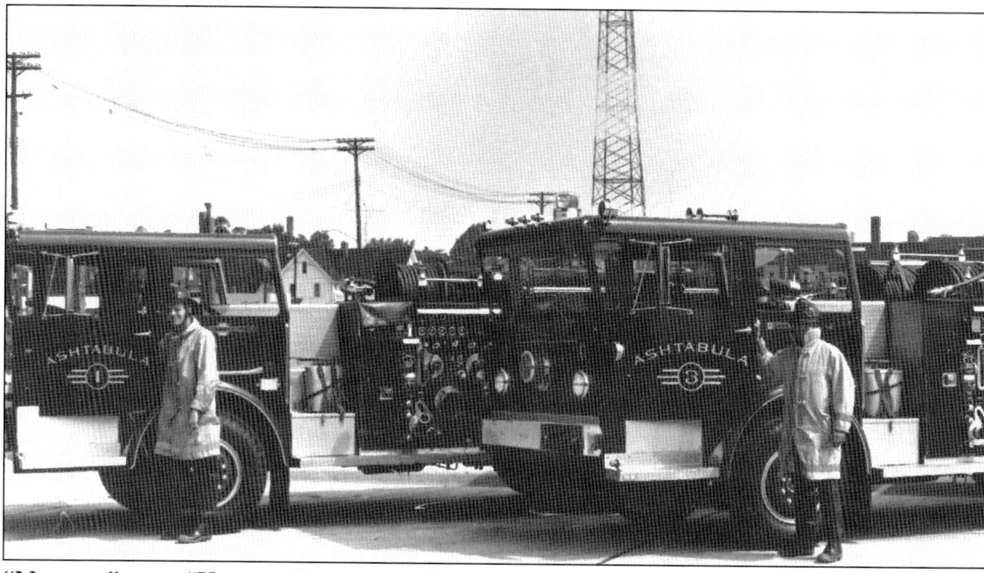


**"NABISCO" AND "KEEBLER," 1975.** Purchased in 1971 by fire chief Charles L. Mosier and city manager George E. Ducro III, these $36,000 and $35,000 factory demonstrator American-LaFrance "Pioneers" pumped 1,250 gallons per minute, a huge improvement over the 750-gallon-per-minute 1944 and 1950 American-LaFrance pumpers they replaced. Both posted to Uptown, the boxy Pioneers were affectionately dubbed "Nabisco" and "Keebler" by firefighters for their resemblance to cracker and cookie boxes. (Courtesy AFD Archive.)

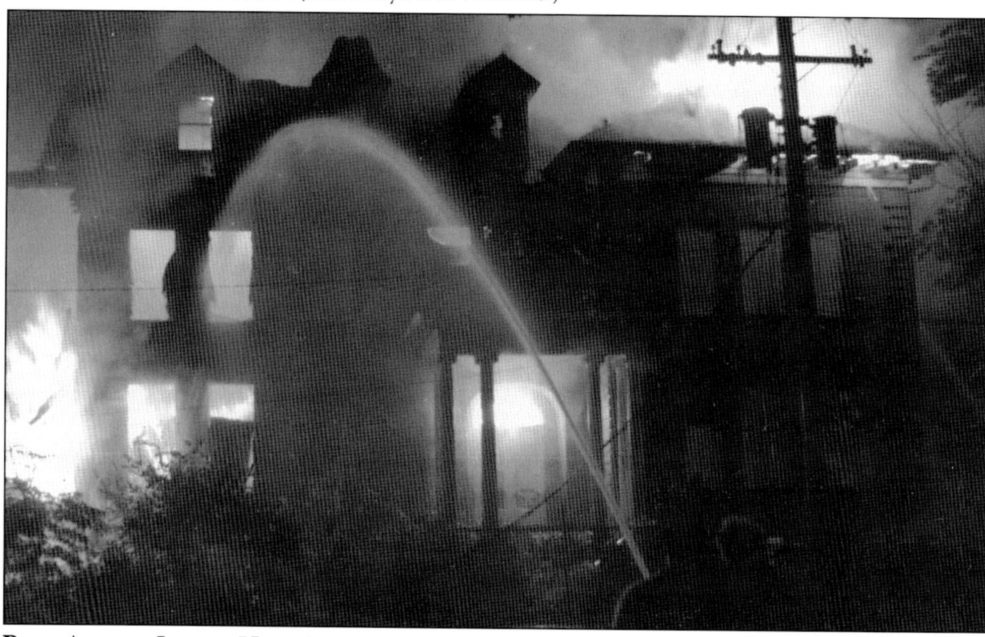

**PARK AVENUE JUNIOR HIGH SCHOOL, 4325 PARK AVENUE—SEPTEMBER 1975.** City firefighters were assisted by Ashtabula, Saybrook, and Plymouth Township firefighters in fighting this $2 million arson fire at the shuttered, National Register–listed Park Avenue Junior High. Home to Ashtabula High School (1900–1916) and Park Avenue Junior High (1916–1957), the picturesque Georgian Revival–styled school later housed Kent State University-Ashtabula (1958–1967). (Courtesy AFD Archive.)

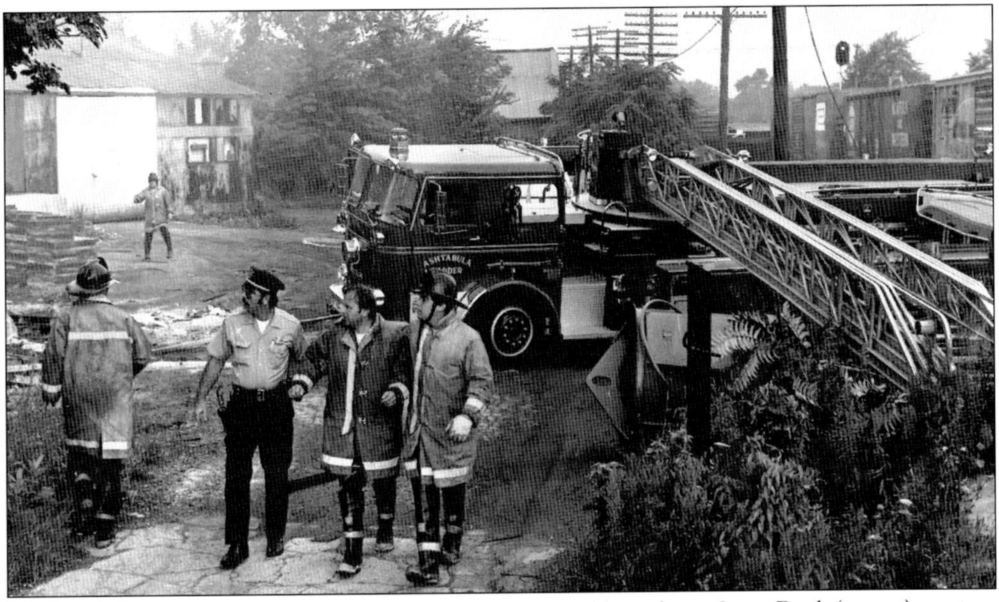

**Ladder Truck Accident, August 14, 1976.** Dazed Ladder 1 driver Steve Deak (center) receives assistance after an eastbound Norfolk and Western freight (upper right) struck the city's *c.* 1966 Seagrave aerial ladder truck during a warehouse fire at Chestnut and West Forty-ninth Streets. The train missed Norfolk and Western's stop call, causing $22,129 in damage to 100-foot Ladder 1, the largest aerial truck between Painesville and Erie. (Photograph Ashtabula Star Beacon; courtesy AFD Archive.)

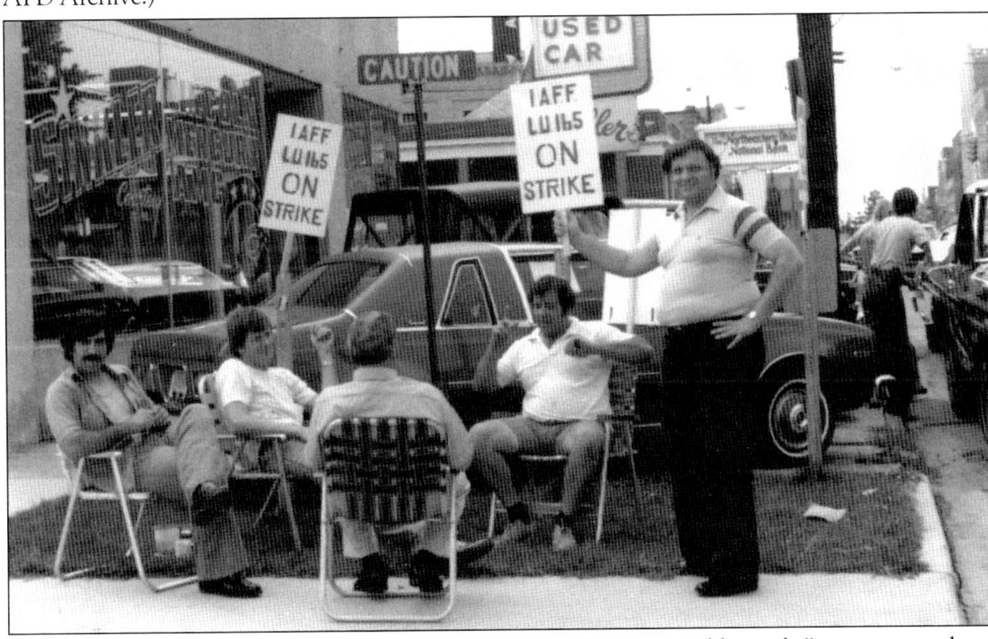


**IAFF Local 165 Strike, September 1–2, 1977.** After "no appreciable results" in nine months of negotiations after their contract expired, 36 IAFF Local 165 firefighters went on strike. A one-year compromise contract called for a $620 wage increase in 1977 and a two-hour reduction to a 54-hour week in 1978. A state law later barred Ohio's police and fire employees from striking, sending labor disputes to binding arbitration. (Courtesy AFD Archive.)

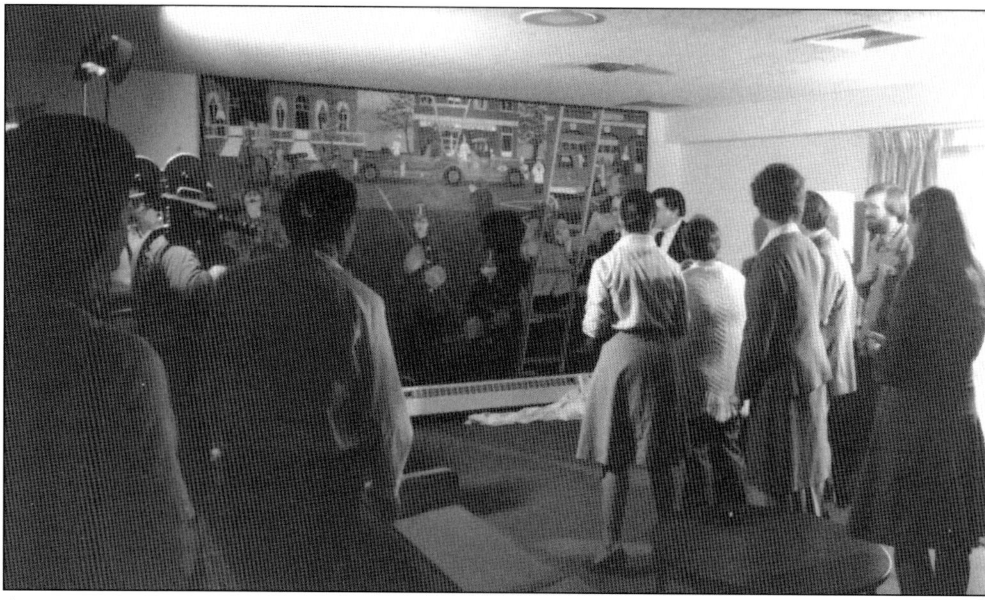

**Ashtabula Firefighting Mural, April 1978.** Station One's day room fills as popular WKYC-TV 3 reporter Del Donahoo films a feature on artist Bob Dasher's enduring 7-by-12-foot Ashtabula Fire Department history mural. Dasher, employed by the county to paint the mural with federal manpower funds, spent four months researching and painting the mural. In 2004, department retiree Harold Leehan hired Dasher to touch-up, update, and expand the mural. (Courtesy AFD Archive.)

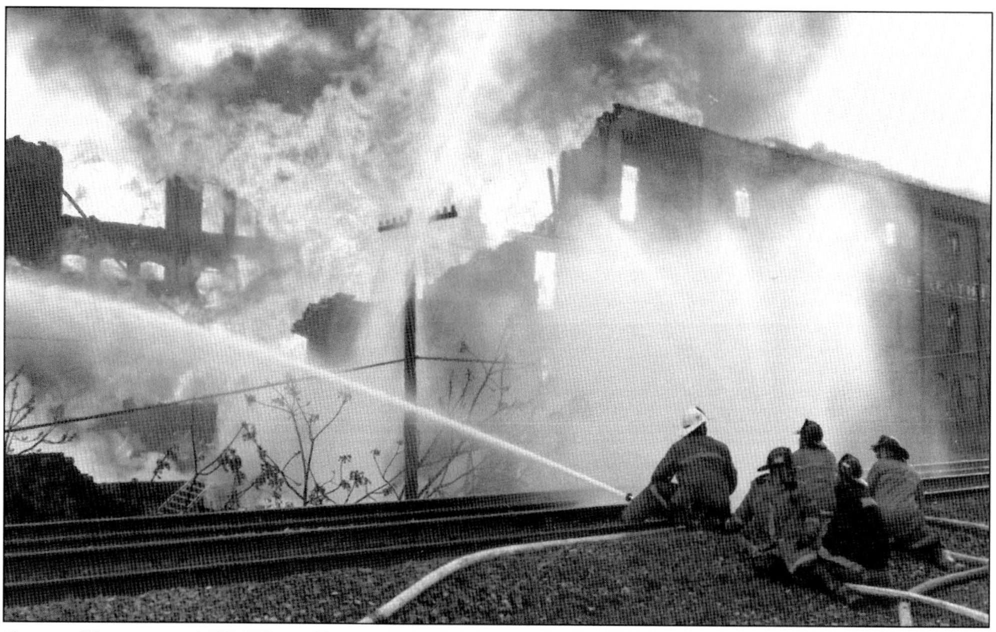

**Raser Tannery, 757 West Prospect Avenue—May 1982.** Ashtabula's abandoned Raser Tannery was destroyed in this $1 million arson blaze, fought by 150 area firefighters. Some 27 firefighters and police officers were injured during the haz-mat blaze, which also damaged two city fire trucks. Vacant Raser was hit by three arson fires between 1980 and 1982. At its peak, Raser Tannery (1896–1980) employed 120. (Courtesy Ashtabula Township Fire Department.)

Six

# THE JEPSON, BALOG, AND PRISTERA ERA

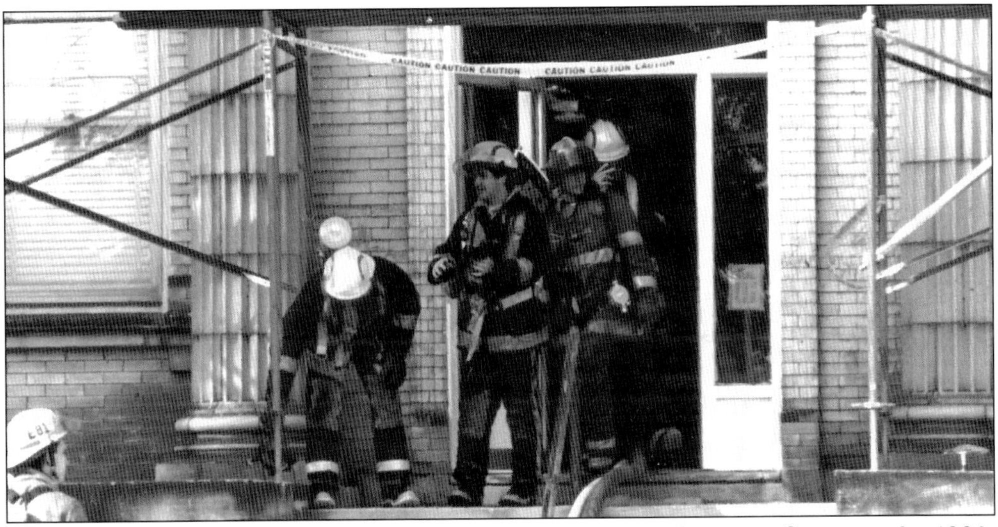

ASHTABULA PUBLIC LIBRARY, 335 WEST FORTY-FOURTH STREET—OCTOBER 3, 1990. Just months shy of completing a $1.2 million renovation, a multimillion-dollar fire struck the Ashtabula Public Library's picturesque *c.* 1903 downtown home, designed by Ashtabula architects Thayer and Wilson and donated by prominent immigrant Scottish industrialist and philanthropist Andrew Carnegie. Over 51 firefighters from Ashtabula City, Ashtabula Township, Geneva, Jefferson, and Saybrook Township battled the blaze. Relocated to the former Carlisle Department Stores office building on State Road, the library returned home in March 1992. With thousands of volumes in its 80,000-piece collection destroyed, including the bulk of its children's and young adults collection, a January 1991 column on the library's plight by nationally syndicated Chicago *Tribune* columnist Bob Greene saw the grateful library deluged with 30,000 books from libraries and individuals across the nation. Established as a subscription library in 1813, the library was reorganized as a free public library in 1895. Built in 1903, the current library building was expanded in 1958 and 1984. (Photograph C. Ronald Greenly; courtesy AFD Archive.)

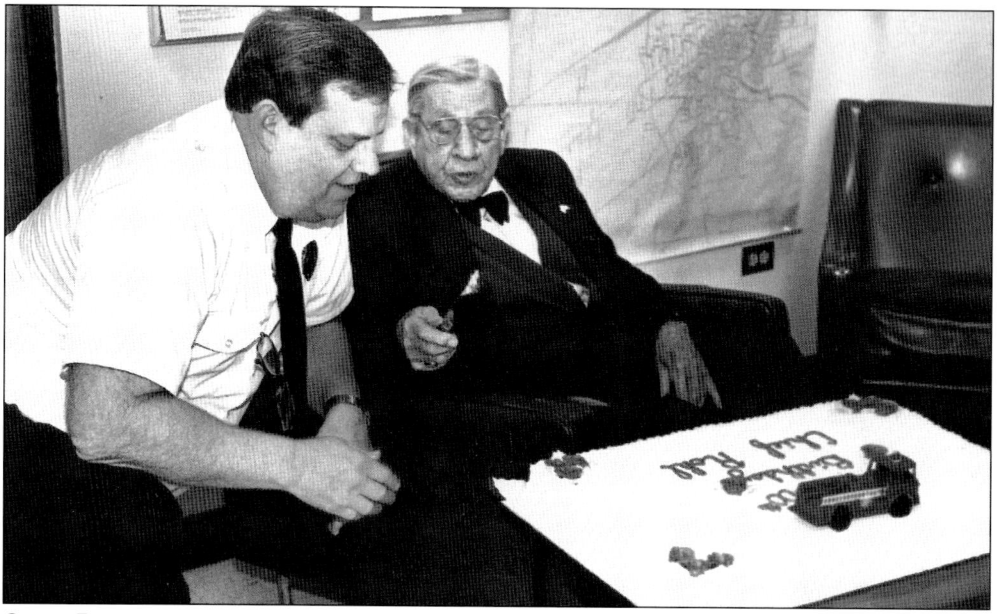

**CHIEF ROHL 100TH BIRTHDAY PARTY, FEBRUARY 1992.** Capt. Charles P. Brockway presents a birthday cake to "living legend" retired 1945–1963 fire chief Adolf T. "Smokey" Rohl on his 100th birthday. Brockway (1965–1998) had been inspired by Rohl as a young boy to enter the fire service. Rohl passed away at age 101 in October 1993. A procession of city fire trucks led Rohl's funeral procession to Edgewood Cemetery. (Courtesy AFD Archive.)

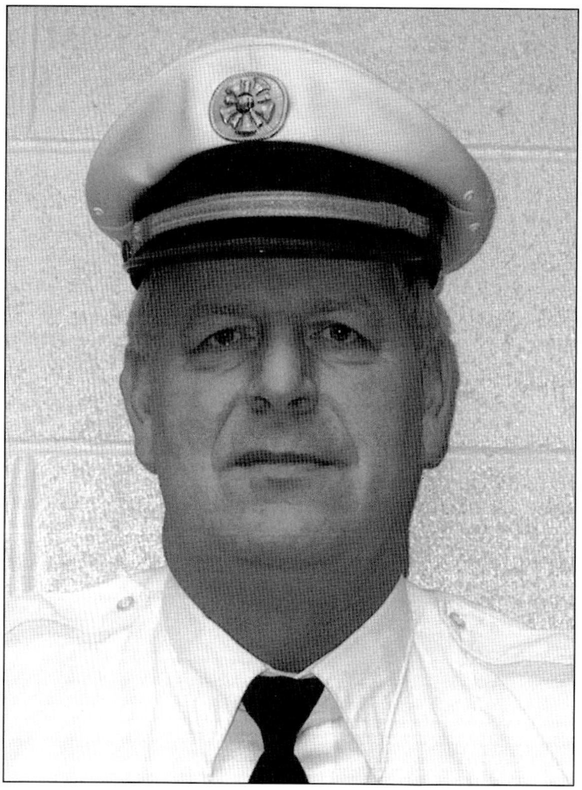

**FIRE CHIEF NORMAN JEPSON.** From 1987 to 1997, the Ashtabula Fire Department was headed by Norman Jepson (1971–1997), promoted to engineer in 1974, lieutenant in 1978, and captain in 1982. Neither Jepson nor predecessor chief Charles Mosier had easy tenures, given the city's financial problems in the 1970s–1990s, a period marked by significant staffing cutbacks, limited department budgets, and the January 1988 closure of the Harbor Fire Station. Tenure highlights included the 1991 addition of the department's 102-foot Grumman Aerialcat ladder truck. (Courtesy AFD Archive.)

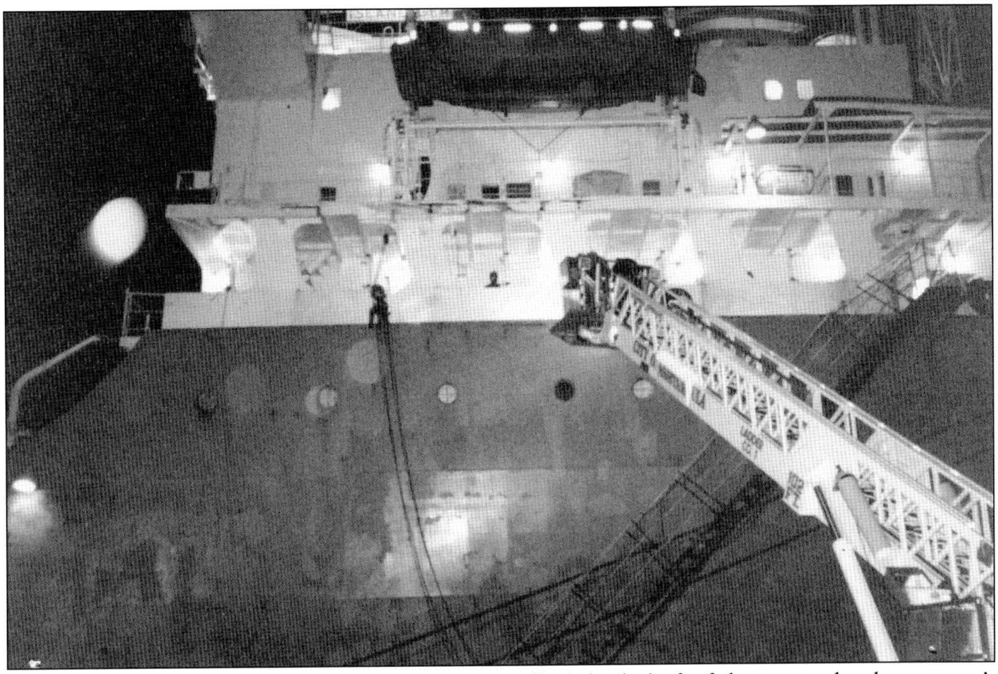

**ISLAND GEM, 1992.** During this evening rescue call, Ashtabula firefighters use the department's new c. 1991 Grumman Aerialcat ladder truck to reach a heart attack victim aboard the Great Lakes freighter *Island Gem*, moored in Ashtabula Harbor. When fully extended, the Grumman's 102-foot ladder can provide access to the upper reaches of most city buildings. (Courtesy AFD Archive.)

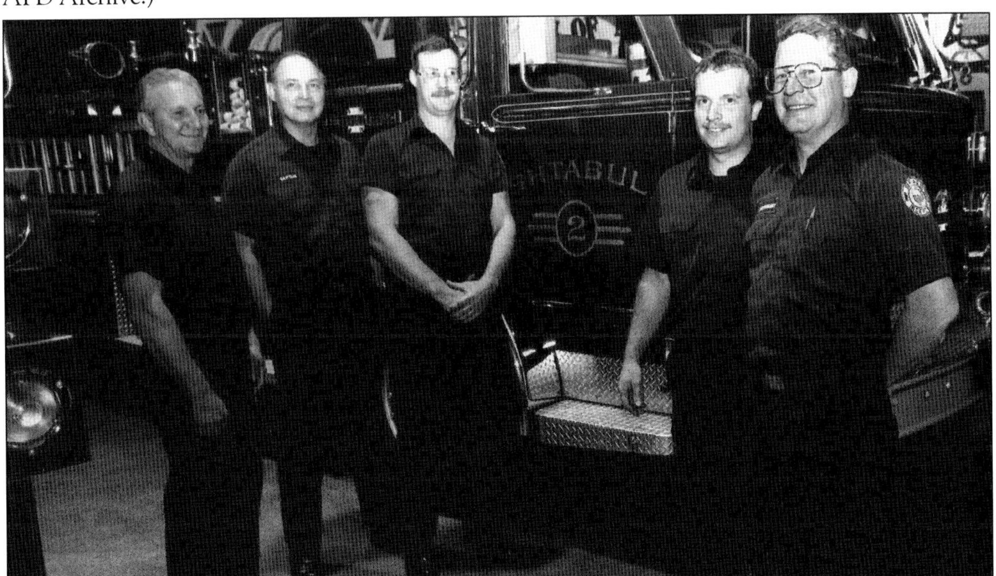

**"HEROES IN THE COMMUNITY," 1993.** From left to right, firefighters Dave Britton, Capt. Lyle G. Hare, Engineer Mike Clayman, Scott Andrus, and Lt. Ross Caudill were feted by the city for rescuing 20 trapped infants, children, and adults at a July 31, 1993, arson blaze at the Coastal Place Apartments, 1211-1217 West Prospect Avenue. Britton, a 27-year department veteran, retired in 2000. (Photograph Warren Dillaway; courtesy AFD Archive.)

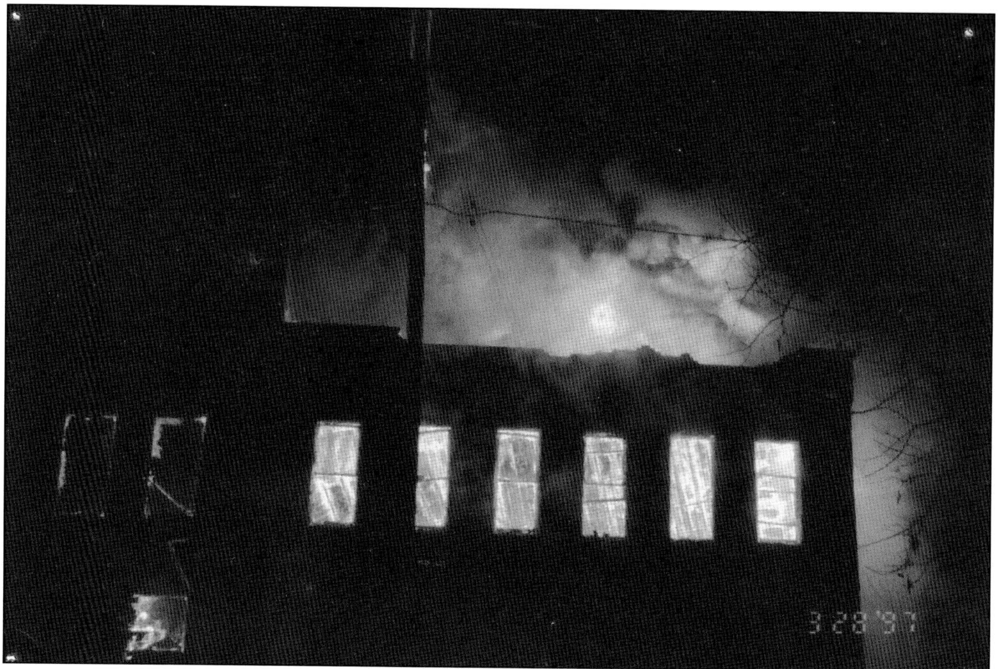


**ASHTABULA BOW SOCKET FIRE, 4212 ANN AVENUE—MARCH 28–29, 1997.** The long-vacant c. 1890 Ashtabula Bow Socket Company factory was gutted in this dramatic arson blaze that "lit the northern skyline for miles." The 10-hour fire was fought by 40 firefighters from Ashtabula City and Saybrook and Ashtabula Townships. Ashtabula Bow Socket (1881–1972) once employed over 100 in the manufacture of drop-forged tools. (Courtesy AFD Archive.)

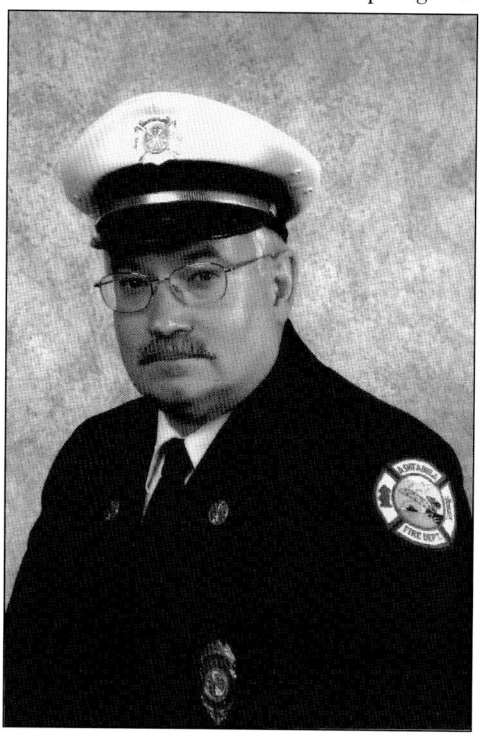

**FIRE CHIEF RICHARD F. BALOG JR.** The son of Ashtabula Police Department patrolman Richard F. Balog Sr., Harbor High School alumnus Rick Balog joined the Ashtabula Fire Department in 1975. Promoted to driver in 1978, lieutenant in 1981, and captain in 1995, Balog was named fire chief in 1997, retiring in March 2006. Tenure highlights include the purchase of the department's c. 1998 HME/General and c. 2006 Sutphen pumpers. (Courtesy Gary Cartner.)

**CHUCK BROCKWAY RETIREMENT, JULY 1998.** A November 1997 knee injury forced the retirement of 33-year department veteran Capt. Charles P. Brockway (1965–1998), a 31-year IAFF Local 165 officer. Brockway logged 25 years with the department's Fire Inspection Bureau, spearheading fire prevention and arson investigation activities. In February 1997, Brockway became a local celebrity with his daring role in helping apprehend two First Merit bank robbers. (Courtesy Charles P. Brockway.)

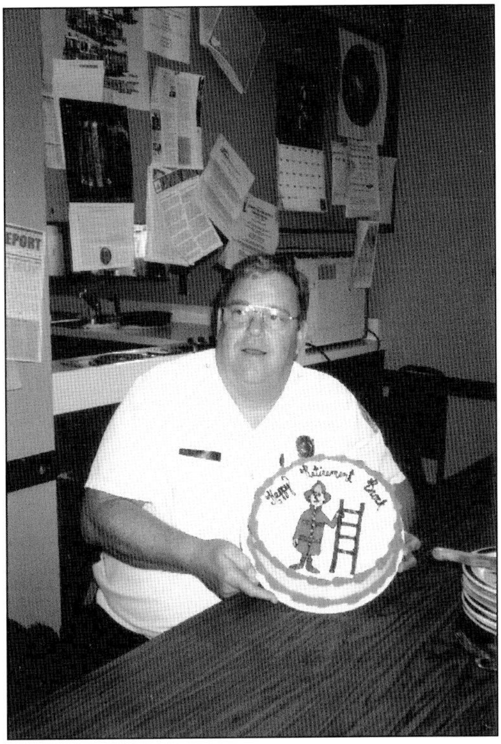

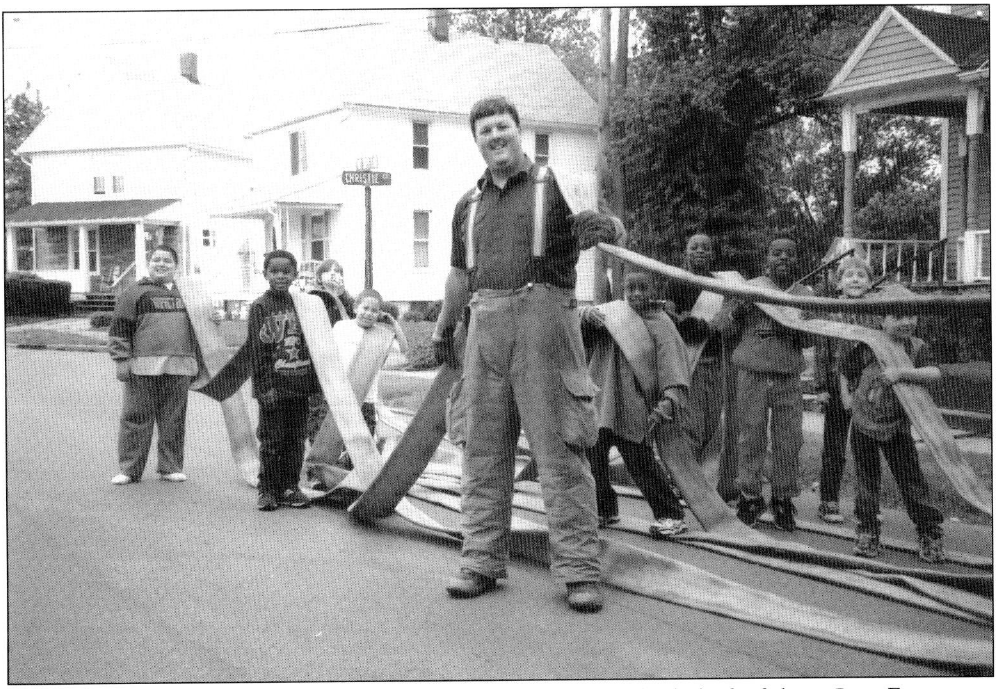


**FIREFIGHTER GREG EVANS AND FRIENDS, LATE 1990S.** Ashtabula firefighter Greg Evans gets an enthusiastic assist with repacking fire hose from a multi-cultural rainbow of eager young friends following a fire call near West Thirty-eighth Street and Christie Court. (Courtesy AFD Archive.)

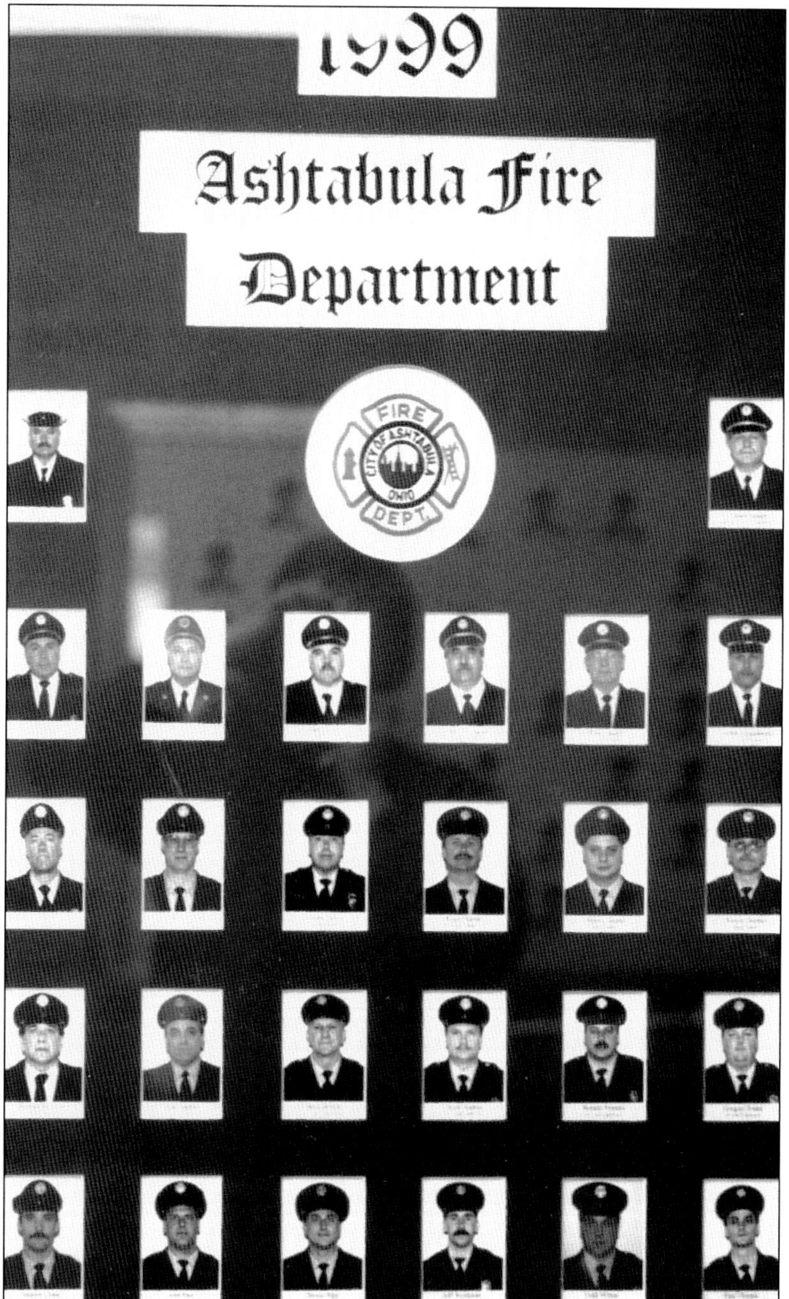


**ASHTABULA FIRE DEPARTMENT ROSTER, 1999.** The department's 26-man, single station 1999 roster is pictured here. They are, from left to right, (first row) firefighters Stephen Chase, John Paul, Dennis Page, Jeff Workman, Todd Wilms, and Paul Thomas; (second row) firefighters Stephen Severino, Alan Guerini, Dave Britton, Scott Andrus, Ronald Pristera, and Gregory Evans; (third row) engineers Kenneth Sweet, Michael Clayman, Mark Dalton, Roger Harrah, Shawn Gruber, and Ronald Deemer; (fourth row) Capt. Thomas Laveck, Capt. Lyle G. Hare, Capt. Mark Calaway, Lt. Randolph Caruso, Lt. Ross Caudill, and Lt. Jeffrey Gianantonio; (fifth row) fire chief Richard F. Balog and Capt. Gerald Senger. (Courtesy AFD Archive.)

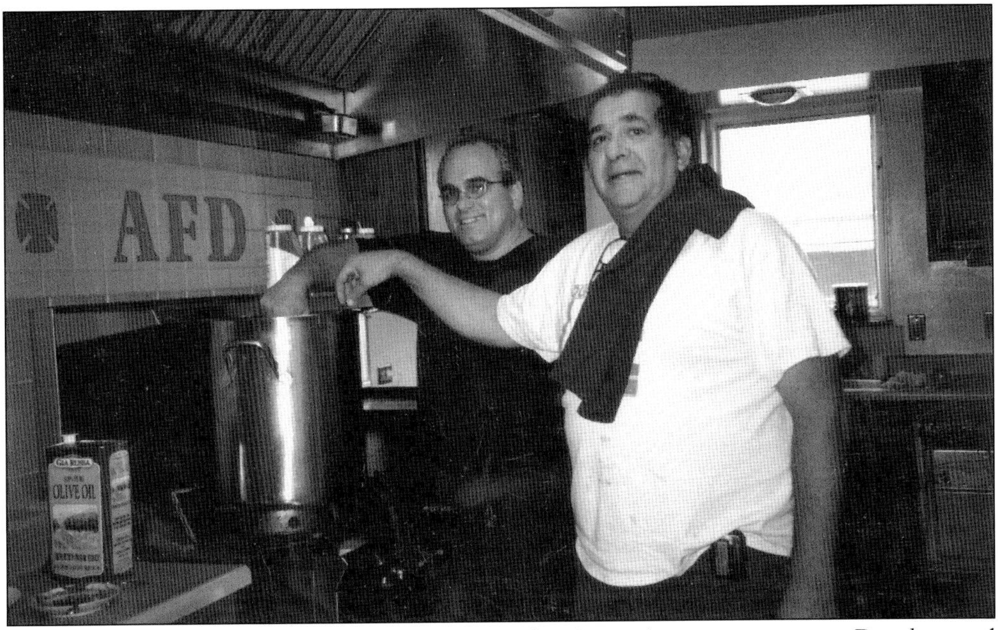


**BON APPETIT.** New York "cooking fireman" John Sineno's appearances on *Donahue* and ABC's *Good Morning America*, coupled with Random House's national 1986 release of Sineno's *Firefighters Cookbook*, brought firehouse cooking into the national spotlight. Notable Ashtabula firefighter cooks include Alan Guerini (left), Steve Severino (right), and Capt. Mark Calaway. A 1990 *Star Beacon* feature showcased Guerini's "Firehouse Cabbage Unrolls" and Calaway's famed beef stroganoff, potato soup, and apple-walnut pie. (Courtesy AFD Archive.)

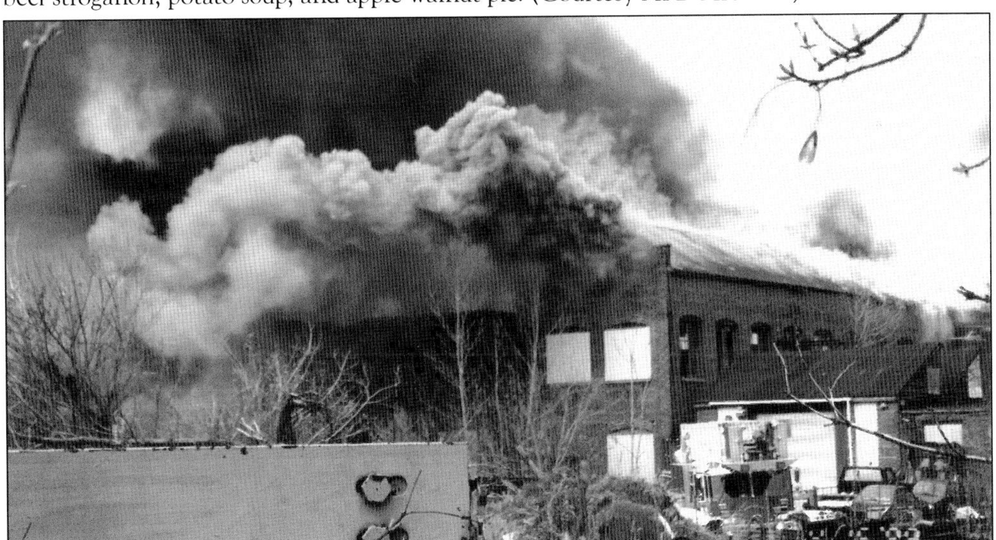

**AMERICAN FORK AND HOE, 409 WEST FIFTY-SEVENTH STREET—APRIL 1999.** This dramatic blaze burned a portion of the former American Fork and Hoe manufacturing complex. Ashtabula firefighters were assisted by Geneva, Saybrook Township, North Kingsville, and Ashtabula Township mutual aid crews. Shuttered in the late 1950s, the "Fork and Hoe," since subdivided and leased to commercial and industrial users, was also struck by fires in 1998 and 2000. (Courtesy AFD Archive.)

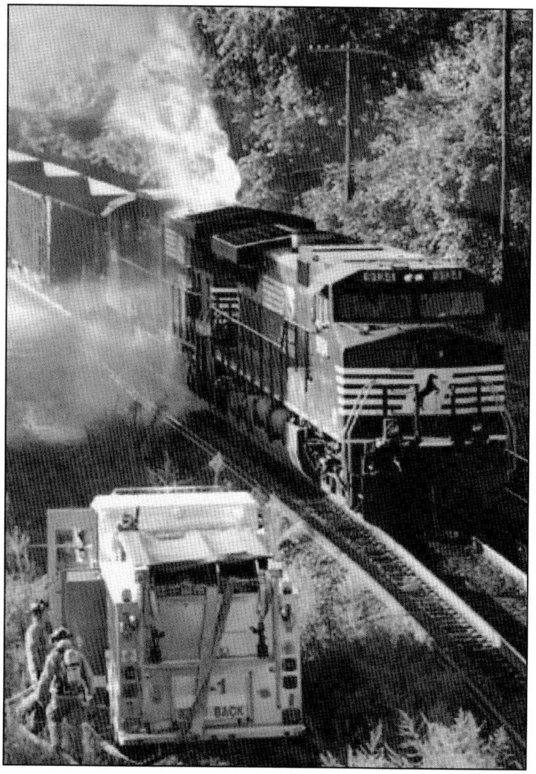

**NORFOLK SOUTHERN ENGINE FIRE—AUGUST 1999.** Shortly after leaving Ashtabula Harbor, this southbound Norfolk Southern locomotive broke out in a dramatic hour-long blaze just north of the U.S. 20 overpass. Loaded with 3,500 gallons of diesel fuel, a 1,200-foot radius around the locomotive was cordoned off, forcing the evacuation of East Prospect Circle homes and the closure of U.S. 20/Prospect Avenue between Station Avenue and Center Street. (Courtesy AFD Archive.)

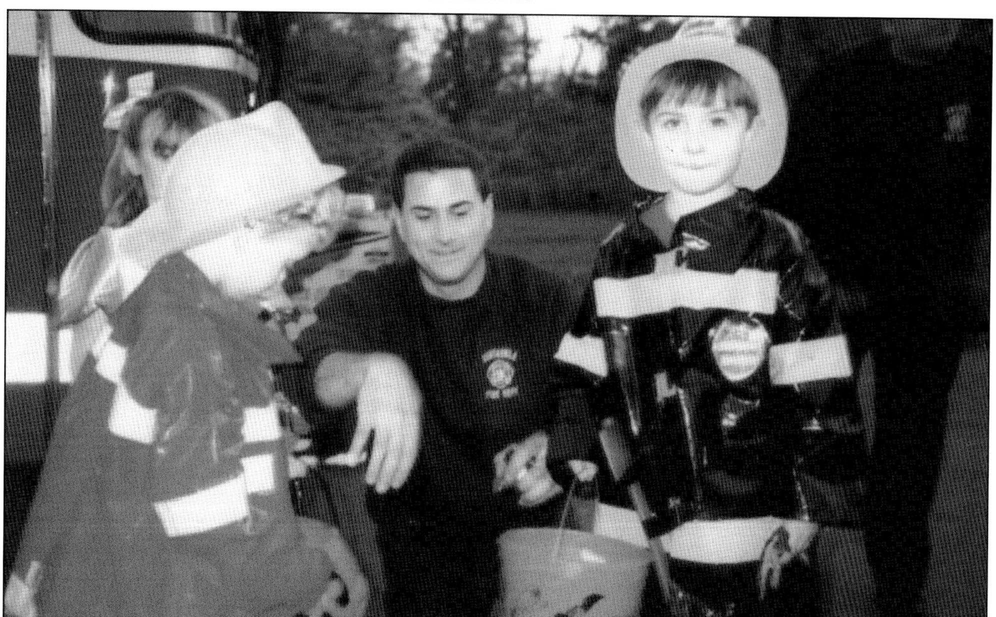

**TRICK OR TREAT, OCTOBER 1999.** Participating in a long-standing Halloween tradition for city firefighters and their union, IAFF Local 165, Ashtabula firefighter Denny Page hands out candy to city youngsters. Financial donations from IAFF Local 165 and the Ashtabula Fire Department's petty cash fund are doubled by civic-minded area stores to purchase Halloween candy for the annual community outreach initiative. (Courtesy AFD Archive.)

HOMETOWN PRIDE, 2000. Because their longtime badge supplier was based out of Cleveland, Ashtabula firefighters long wore patches nearly identical to those worn by Cleveland firefighters, right down to the Cleveland skyline. Seeking a unique identity, fire chief Richard F. Balog Jr. (1997–2006) spearheaded the creation of a localized patch, which includes scenes of Ashtabula Harbor and the city's signature c. 1925 Bridge Street Bascule Lift Bridge. (Courtesy AFD Archive.)

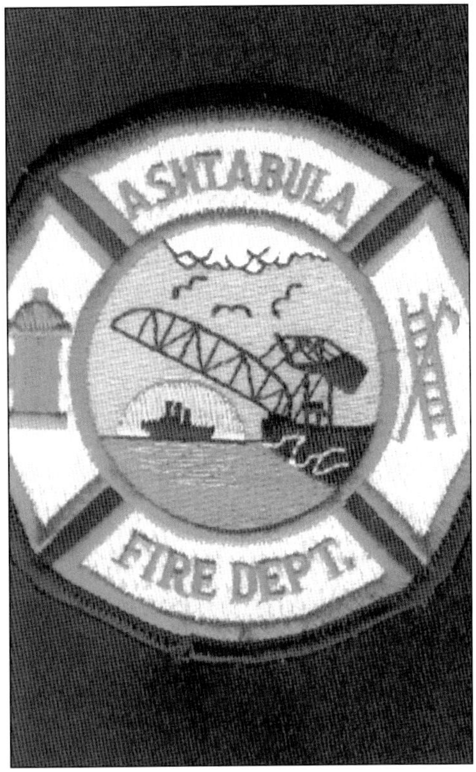


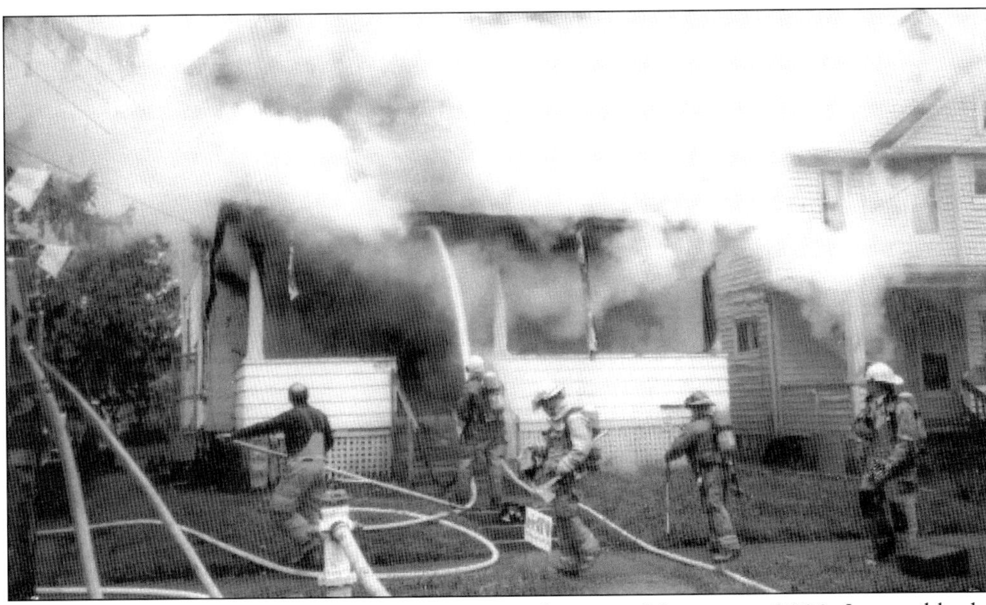

NEWMAN ARSON FIRE, 939 WEST FORTY-THIRD STREET—NOVEMBER 2000. In possibly the worst multi-fatality house fire in city history, this "very gruesome" arson fire claimed the lives of 22-year-old Stephanie Newman, her two- and three-year-old sons, and her 16-year-old sister. Called a "heinous crime," Newman's husband, Charles Newman Sr., entered an Alford Plea and was sentenced on four counts of aggravated murder and four counts of aggravated arson. (Courtesy AFD Archive.)

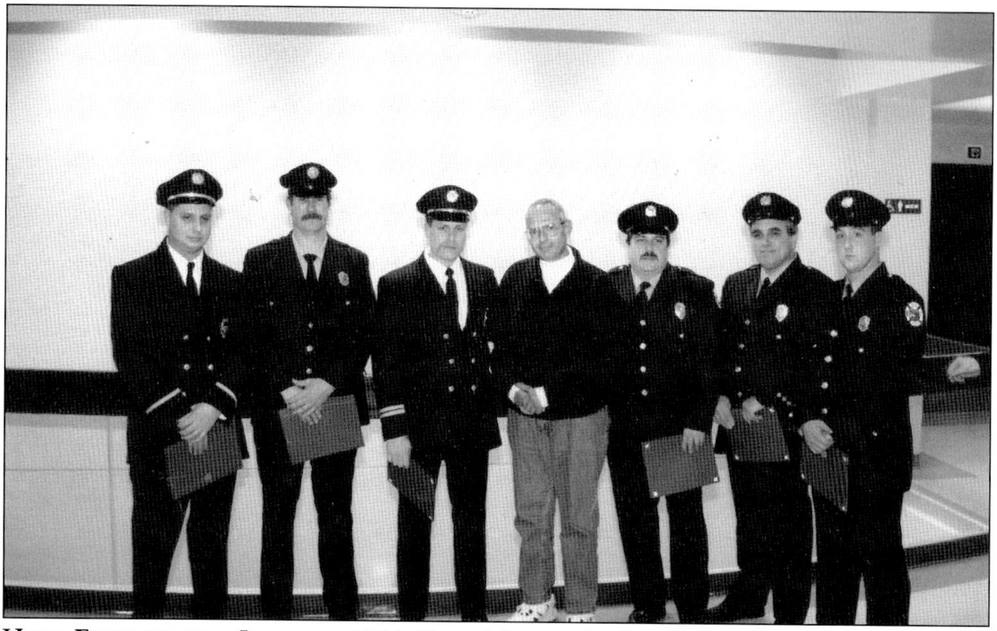

**Hero Firefighters, January 2001.** From left to right, Lt. Shawn Gruber, engineer Mike Clayman, Capt. Jeff Gianantonio, engineer Ronald Pristera, firefighter Alan Guerini, and firefighter paramedic Mike Discher were feted by the city for saving the life of East Fifty-first Street resident Pasquale Silva (center) following a December 2000 natural gas explosion. Suffering burns to his airway, face, neck and hands, Silva was treated at Cleveland-based MetroHealth Medical Center's burn unit. (Courtesy AFD Archive.)

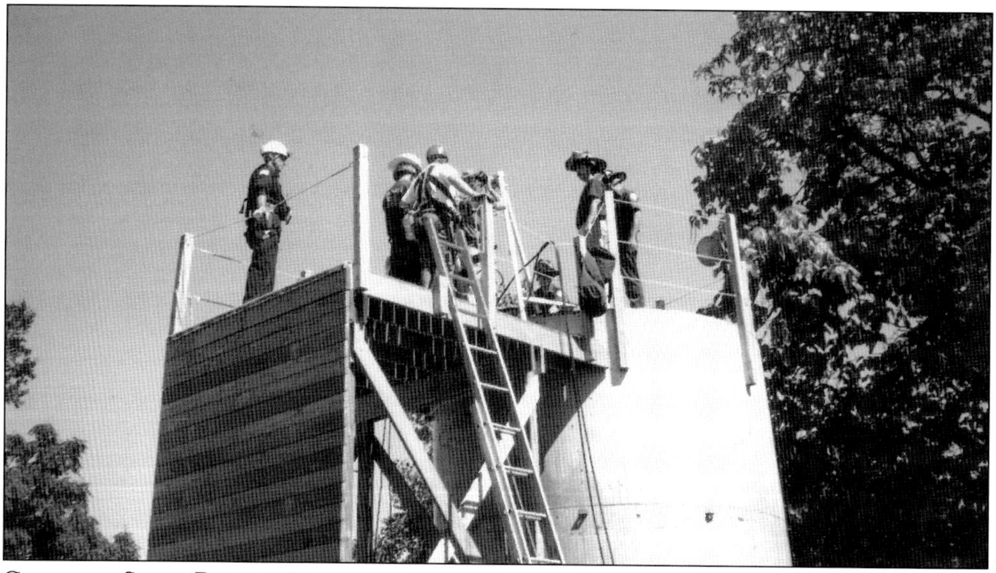

**Confined-Space Rescue Training, 2001.** Keeping their skills sharp, Ashtabula firefighters spend hundreds of hours annually in a variety of refresher courses and hands-on training sessions. Here firefighters utilize Station One's confined-space rescue and rope/rappelling training tower, developed in June 1999 following a donation by Ashland Chemical. Confined-space training familiarizes firefighters with the unique hazards and challenges of confined-space rescues. (Courtesy AFD Archive.)

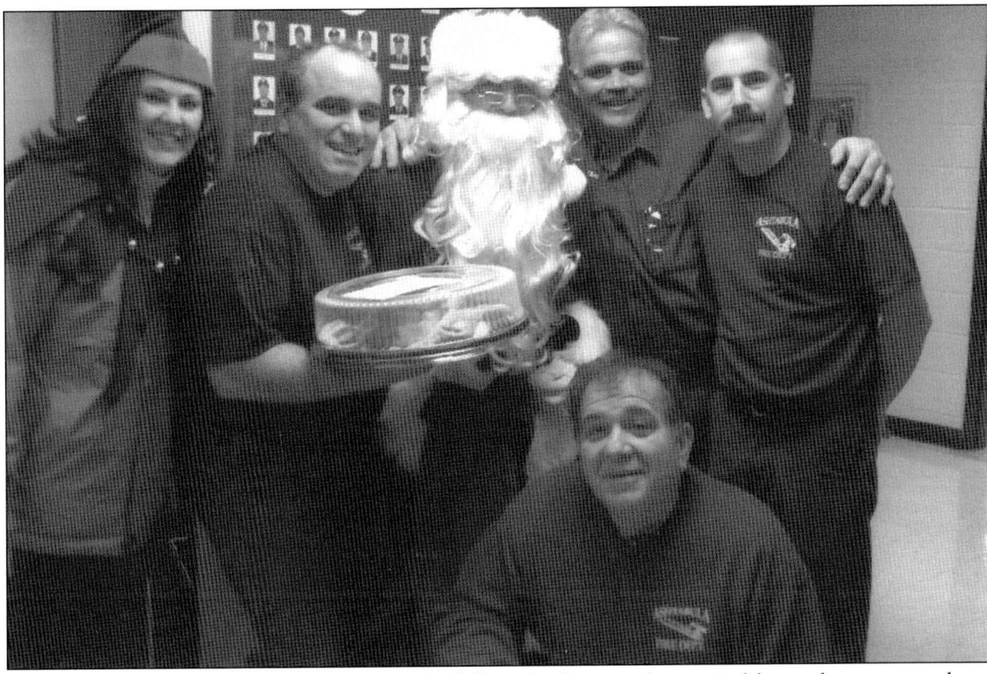

SANTA VISITS STATION ONE, 2001. As folks who know what it is like to have to work on Christmas Eve, away from home and hearth, Santa Claus and an elfin helper make their annual Christmas Eve cookie delivery to on-duty Station One firefighters. Clockwise from left are firefighter Alan Guerini, Capt. Mark Calaway, and firefighters Jeff Workman and Steve Severino. (Courtesy AFD Archive.)

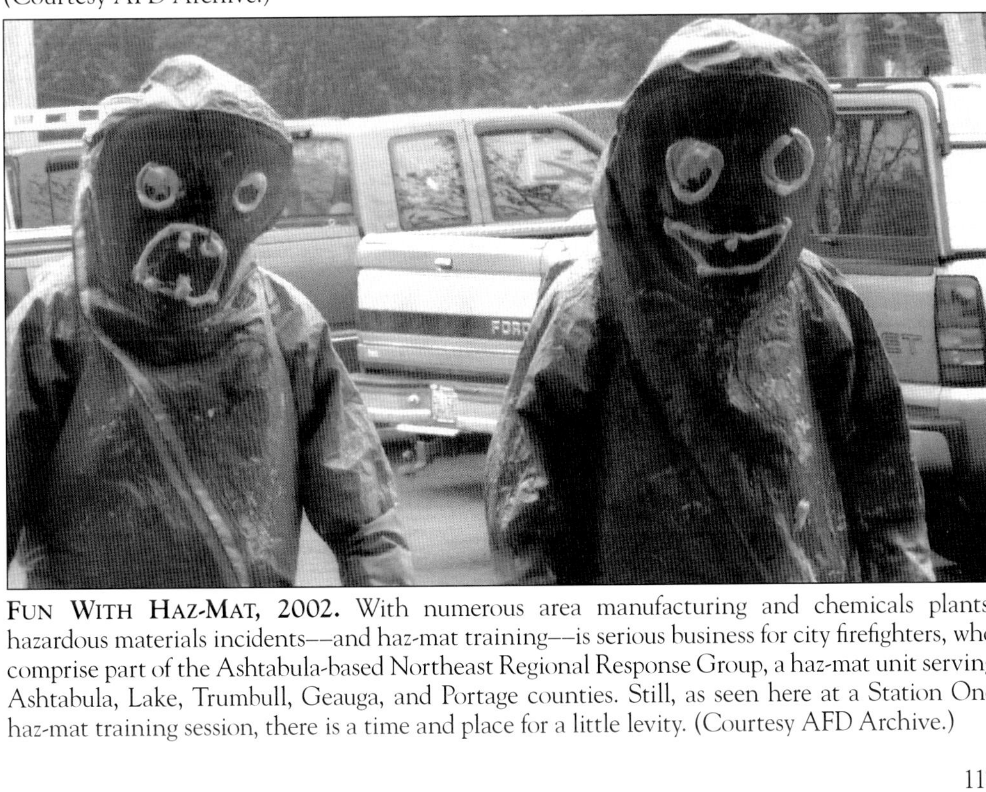

FUN WITH HAZ-MAT, 2002. With numerous area manufacturing and chemicals plants, hazardous materials incidents––and haz-mat training––is serious business for city firefighters, who comprise part of the Ashtabula-based Northeast Regional Response Group, a haz-mat unit serving Ashtabula, Lake, Trumbull, Geauga, and Portage counties. Still, as seen here at a Station One haz-mat training session, there is a time and place for a little levity. (Courtesy AFD Archive.)

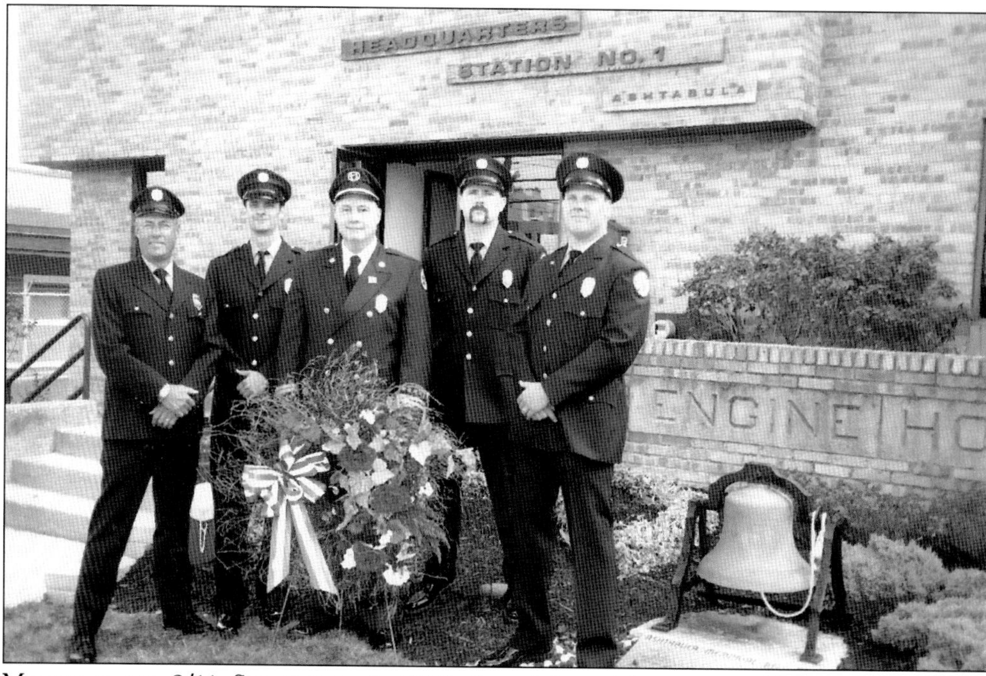


**MEMORIAL FOR 9/11, SEPTEMBER 11, 2002.** From left to right, engineer Ken Sweet, firefighter/paramedic Paul Thomas, Capt. Lyle G. Hare, firefighter Doug Roxberry, and engineer John Paul gather outside Station One for a ceremony honoring the 2,986 victims of the September 11, 2001, terrorist attacks on New York's Twin Towers, the Pentagon in Washington, D.C., and United Flight 93, including 343 City of New York Fire Department "brothers." (Courtesy AFD Archive.)

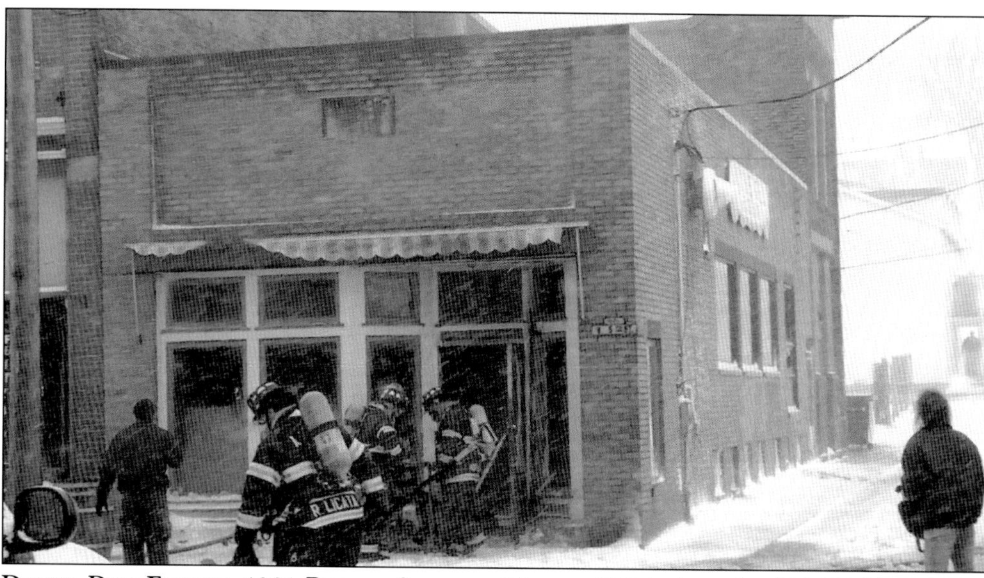

**DOXSIE DELI EATERY, 1001 BRIDGE STREET—MARCH 2004.** A smoky $90,000 blaze forced the temporary closure of Doxsie Deli Eatery (1991– ), a popular Ashtabula Harbor riverfront restaurant famed for its dramatic views of the city's historic c. 1925 Bridge Street Bascule Lift Bridge. Doxsie Deli, reopened in March 2005, takes its name from its location in the c. 1935 Doxsie Dairy Building. (Courtesy AFD Archive.)

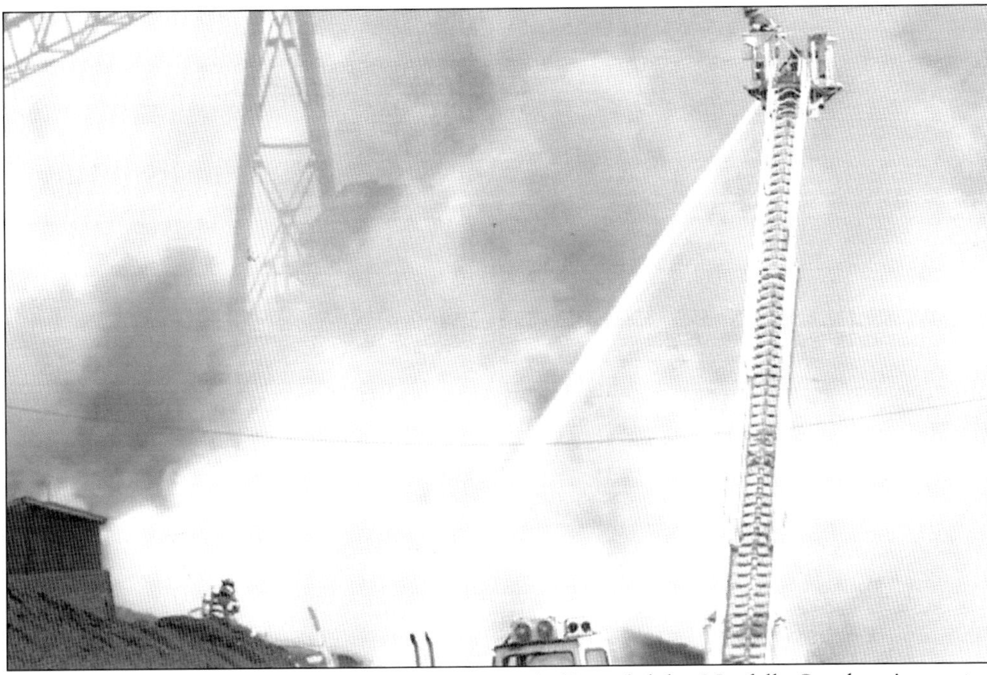

**Norfolk Southern Coal Dock, June 2005.** Dwarfed by Norfolk Southern's soaring c. 1968 Ashtabula River coal conveyor bridge (top left), the Ashtabula Fire Department's 102-foot Grumman Aerialcat sprays water onto a burning maintenance building at Norfolk Southern's Ashtabula Harbor coal dock. Six engines and two ladder trucks staffed by 37 Ashtabula City and Ashtabula, Kingsville, and Saybrook Township firefighters fought the five-hour blaze and safeguarded an electrical substation. (Courtesy AFD Archive.)

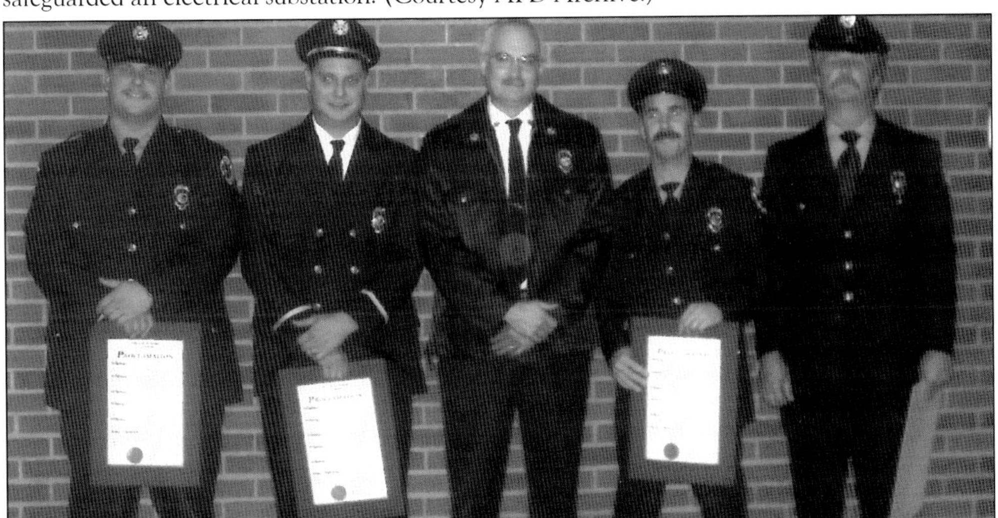

**Hurricane Katrina Relief Proclamations, November 2005.** Matt Burgett, Lt. Shawn Gruber, Jeff Workman, and engineer Roger Harrah, pictured with fire chief Richard Balog (center), were feted by city council for their volunteer Emergency Management Assistance Compact (EMAC) deployment east of New Orleans in hard-hit St. Bernard Parish, a 66,129-resident Louisiana county devastated by Hurricane Katrina. Other county fire departments participating in EMAC deployments were Geneva and Ashtabula Township. (Courtesy AFD Archive.)

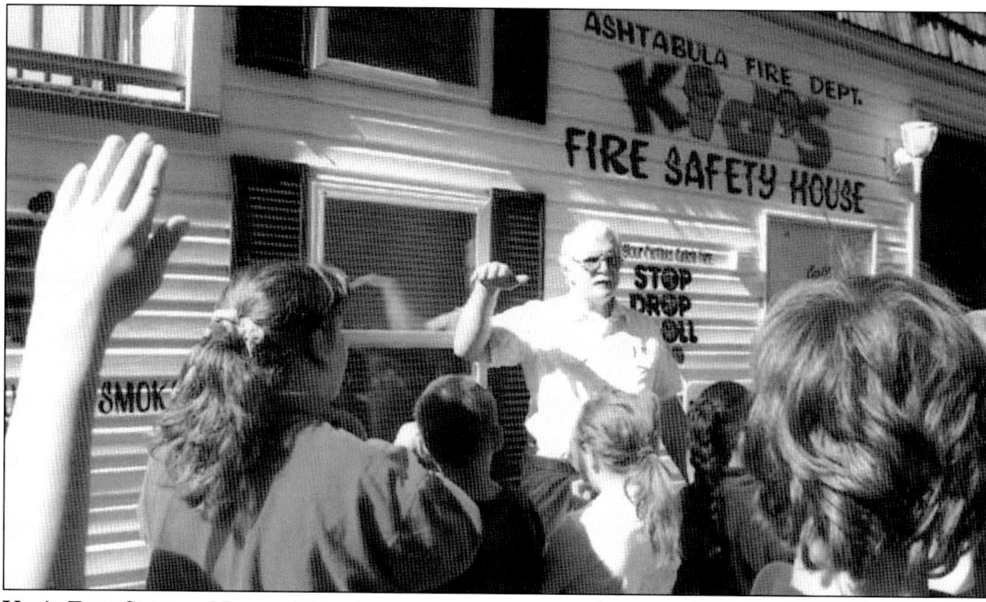


**Kid's Fire Safety House.** Bringing a proactive approach to city firefighting in 1967, fire chief James Doran established the department's Fire Prevention Bureau, headed by full-time fire inspectors Capt. Harry McClimans (1967–1973), Capt. Anthony Candela (1973–1976), Capt. Charles P. (Chuck) Brockway (1976–1998), and Capt. Gerald Senger (1998– ). Here 29-year department veteran Senger addresses a school group outside the department's *c.* 1995 Kid's Fire Safety House fire escape simulator. (Courtesy AFD Archive.)

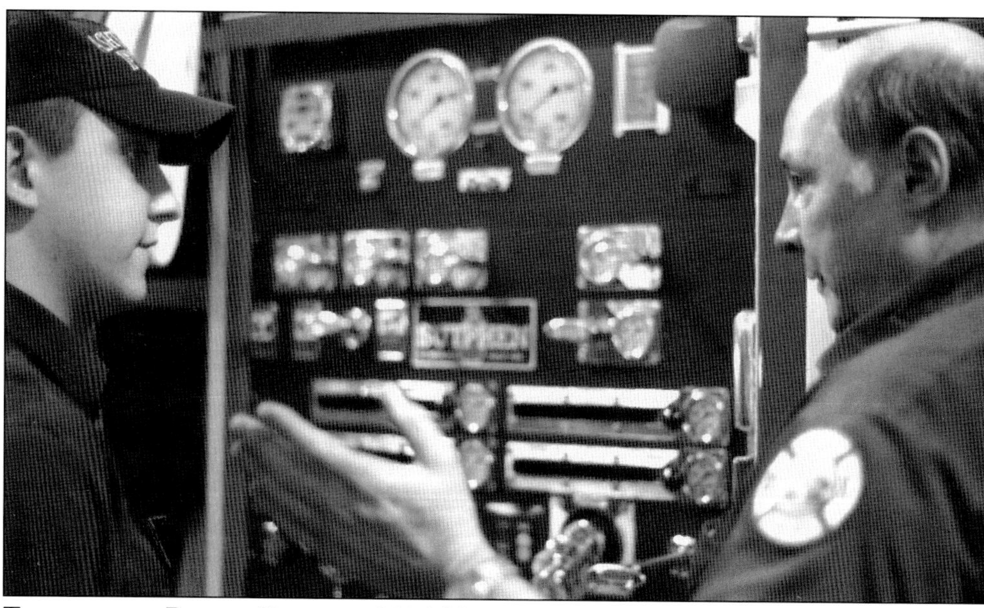

**Training the Rookie, February 2006.** Veteran 37-year firefighter Capt. Lyle G. Hare (right) trains probationary firefighter Geoff Cannon on the operation of the department's new 2006 Sutphen rescue pumper. A Conneaut High School alumnus and 1966–1968 U.S. Army veteran, Hare joined the department in 1969, promoted to driver in 1971, lieutenant in 1976, and captain in 1978. Hare and Cannon are among the nation's 275,000 career firefighters. (Author's collection.)

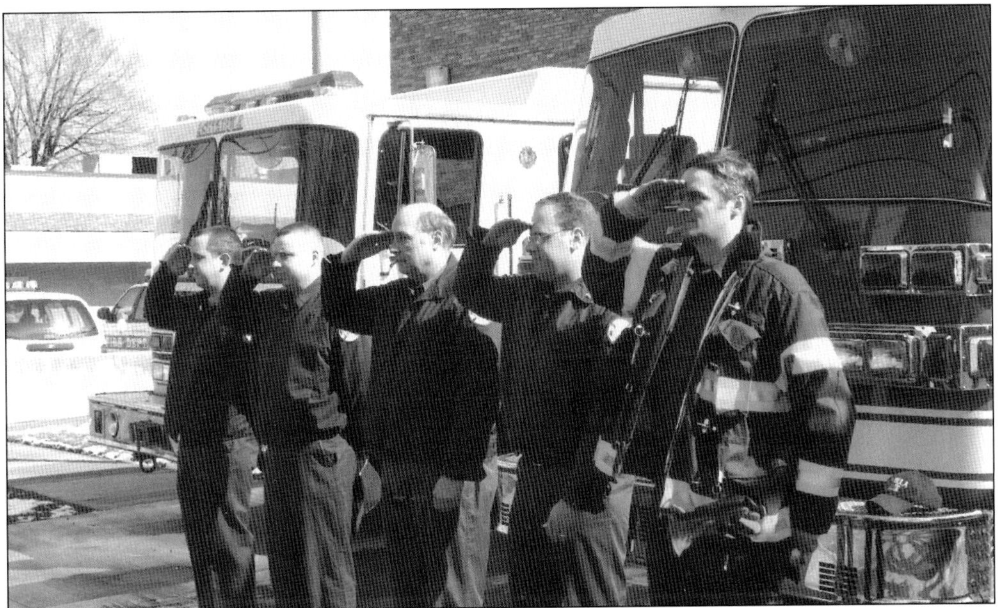

**Saluting a "Brother," February 2006.** With Station One's flag flying at half mast, C-Shift firefighters Geoff Cannon, Richard Hoplight, Capt. Lyle G. Hare, engineer John Paul, and firefighter/paramedic Paul Thomas, from left to right, gather along Main Avenue to pay their final respects as a funeral cortege passes bearing Plymouth Township resident and longtime IAFF union member Eugene B. "Woj" Wojciechowski, a retired 27-year veteran of the City of Wickliffe Fire Department. (Author's collection.)

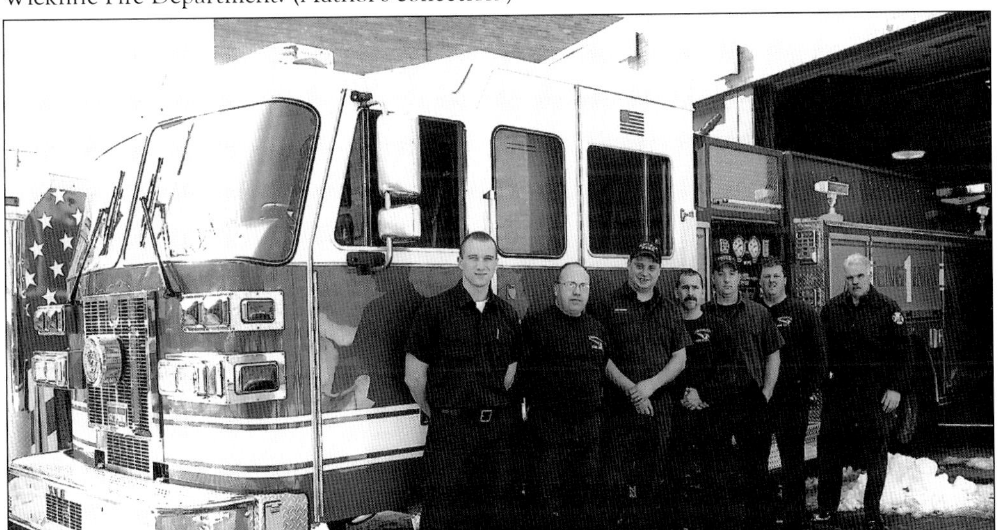

**Sutphen "Monarch" Delivery, February 2006.** From left to right, Joe Swigunski, engineer Mark Dalton, Lt. Shawn Gruber, Jeff Workman, Mike Discher, Greg Evans, and Capt. Mark Calaway take delivery of the department's $350,000 Sutphen "Monarch" rescue pumper. Equipped with a 370-horsepower Cummins ISL diesel engine and a Hale QMAX 1,500-gallon-per-minute pump, Engine 1 was manufactured by Amlin, Ohio's family-owned C. H. Sutphen Corporation (1886– ), the nation's oldest continuously owned-and-operated fire apparatus manufacturer. (Courtesy AFD Archive.)

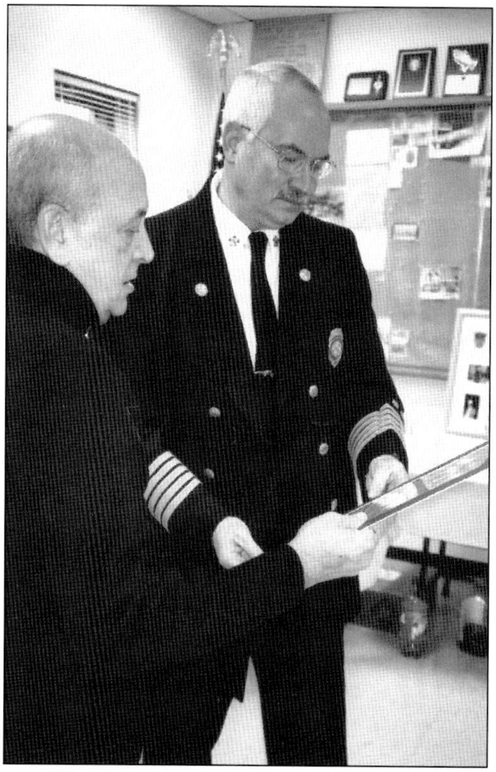

**CHIEF BALOG RETIREMENT PARTY, MARCH 2006.** City manager Anthony J. (Tony) Cantagallo (left) presents retiring nine-year fire chief Richard F. Balog Jr. with a proclamation feting him for his 31-year "devotion to his profession" and the city. While firefighting proved "a good career," Balog noted the profession had hidden costs: "Nobody remembers the good things in this job. The ones you can't forget are the ones you couldn't help." (Author's collection.)

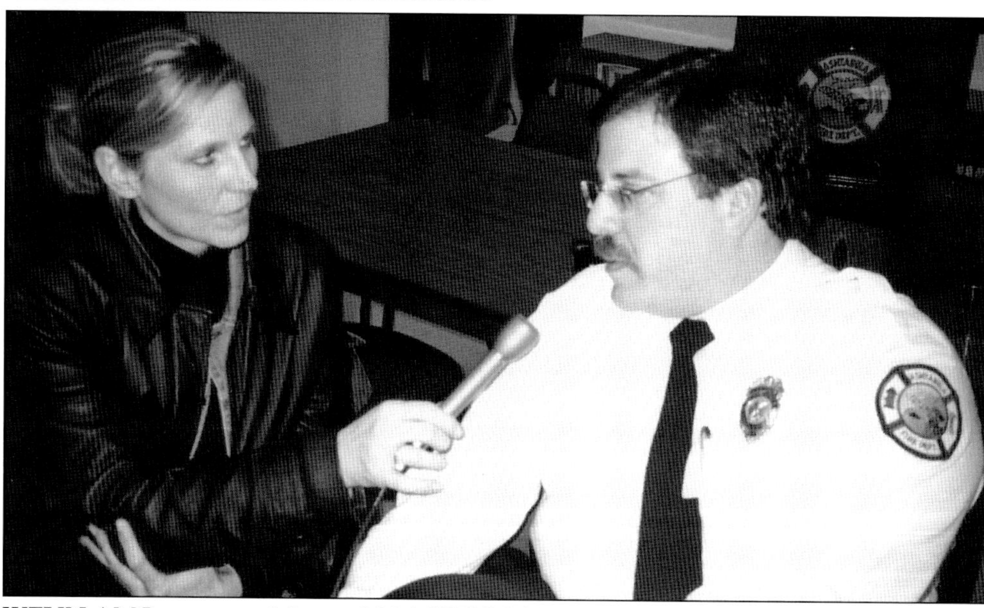

**WFUN-AM INTERVIEW, MARCH 2006.** WFUN-AM 970 news director Kathy Davis interviews incoming Ashtabula fire chief Ronald Pristera. A native of Braddock, Pennsylvania, Pristera joined the department as a firefighter in 1995, promoted to engineer in 2000, lieutenant in 2003, and chief in March 2006. A former 10-year Geneva-on-the-Lake Fire Department volunteer, Pristera concurrently enjoyed a 1985–1995 career as an Ashtabula EMS technician with Ambulance Care and Community Care Ambulance. (Author's collection.)

# *Seven*

# Preserving Department History

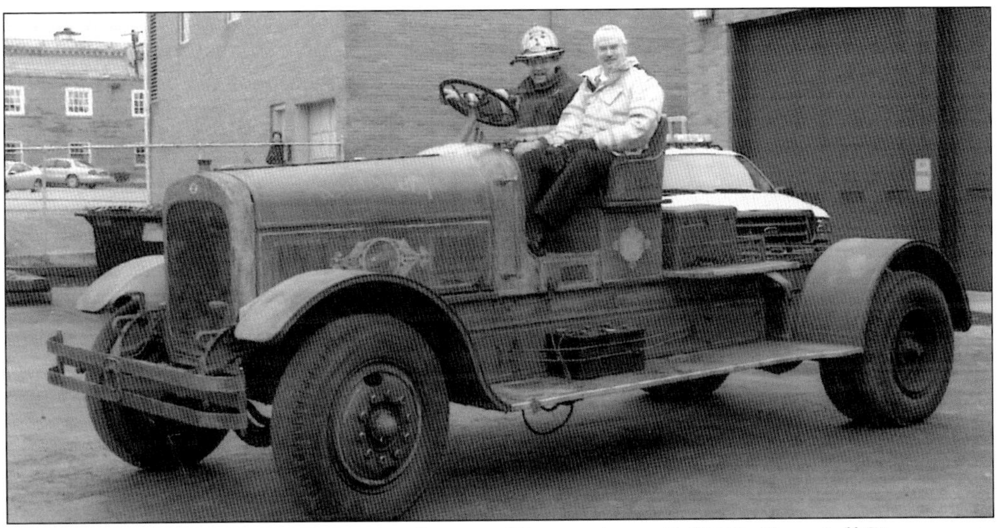

**A 1930 Seagrave Aerial Hook and Ladder Truck, April 9, 2003.** Capt. Jeff Gianantonio (left) and fire chief Richard F. Balog Jr. prepare to drive the propulsion "tractor" section of the Ashtabula Fire Department's old 1930 Seagrave aerial ladder truck. Mustered out of service in 1966 as an antiquated piece of firefighting apparatus, the 75-foot Seagrave was sold to longtime owner Harold Osten of Ashtabula Township. In April 2003, Osten donated the locally significant Seagrave to the Fire Historical Society of Ashtabula, a non-profit corporation organized by preservation-minded Ashtabula firefighters to collect and preserve department artifacts. The group hopes to raise the more than $15,000 needed to return the "very restorable" 1930 Seagrave to its original condition, which included "incredible" hand-painted gold leaf scrollwork and ornate hood paintings of local maritime scenes, including the c. 1906 car ferry *Ashtabula*. Long housed at the old Uptown Fire Station on Park Avenue in downtown Ashtabula, the Seagrave aerial truck was purchased new in 1930 for $18,000, less a $200 trade-in on Moore Hook and Ladder's late 1800s horse-drawn ladder truck. (Courtesy AFD Archive.)

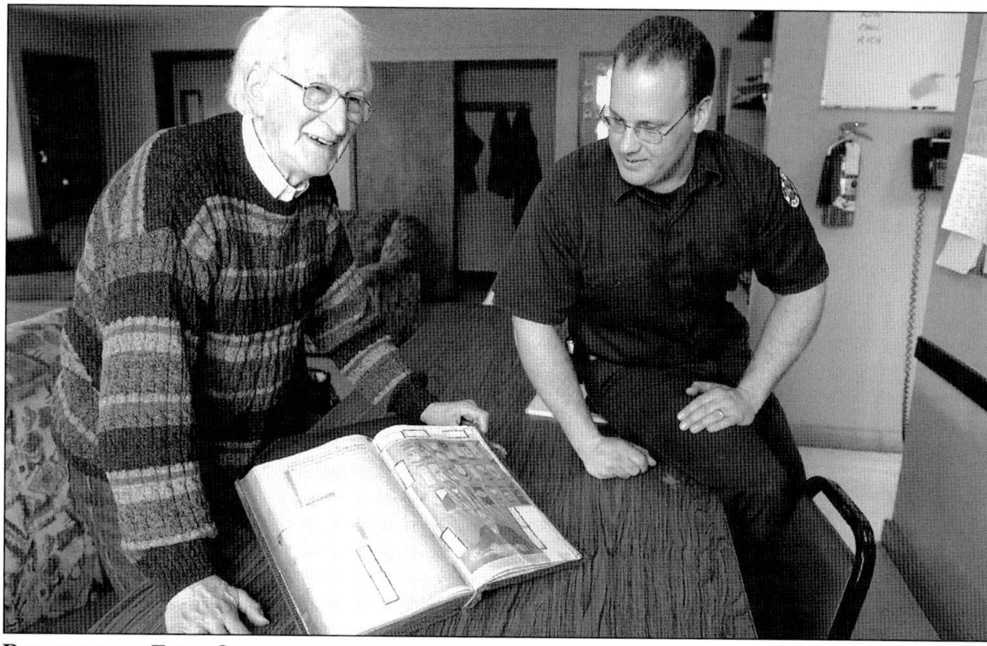

**PROTECTION FIRE COMPANY JOURNAL, FEBRUARY 2006.** As Ashtabula Fire Department engineer and avid historian John Paul (right) looks on, retired Painesville physician Dr. Joseph Koelliker donates a leather-bound 1871–1918 journal chronicling the history of Ashtabula's volunteer Protection Fire Company. Long safeguarded by Protection secretary-treasurer Alfred Kahne, Koelliker purchased Protection's journal at a mid-1990s estate sale, later donating the volume to the department. (Photograph Carl E. Feather; courtesy Ashtabula Star Beacon.)

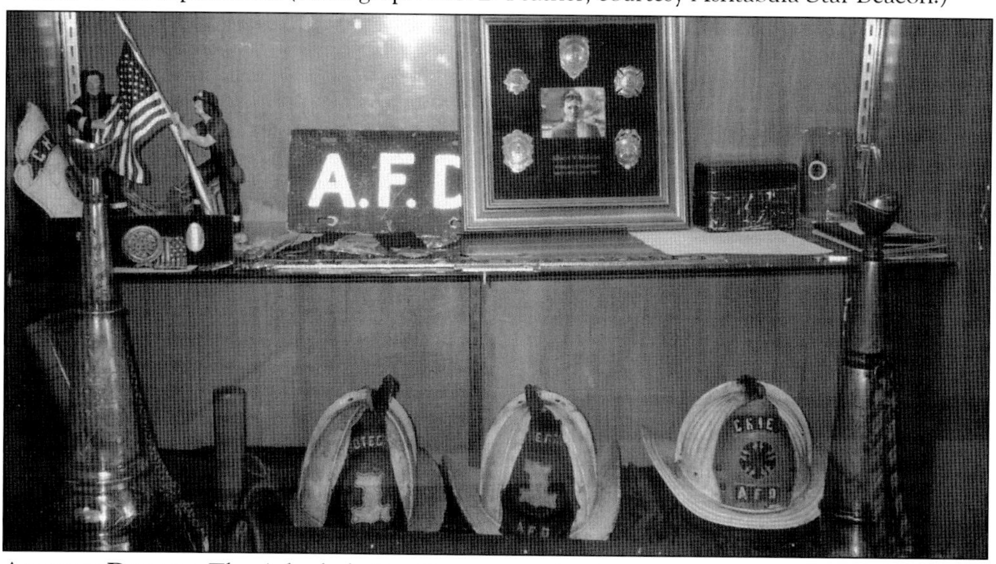


**ARCHIVE DISPLAY.** The Ashtabula Fire Department has been recapturing its rich history with the aid of preservation-minded local history buffs, department retirees, and active firefighters. Historic artifacts include antique fire nozzles, archival photographs, 1880s competition ribbons, vintage leather fire helmets, old log books, and antique badges. Of particular interest are ornate silver-plated "bugle" megaphones from Protection Fire Company (left, 1889) and Neptune Hose Company (right, 1888). (Courtesy AFD Archive.)

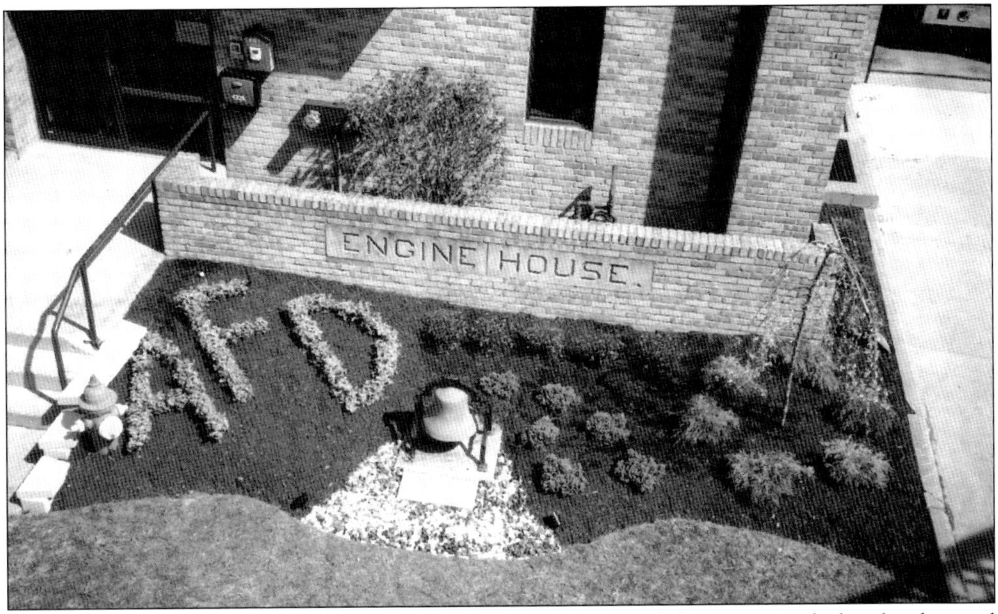


**Ashtabula Bridge Disaster Memorial.** Ashtabula firefighters created this landscaped garden memorializing the 92 victims of the December 1876 Ashtabula Bridge Disaster. Recovery of Lake Erie Hose Company's c. 1854 fire bell, which summoned volunteer firefighters to the tragedy, was coordinated in 1975 by fire chief Charles Mosier and insurance agent Francis Herzog. A granite memorial marker, donated by Smith and Son Monument Company, was dedicated in 1999. (Courtesy AFD Archive.)

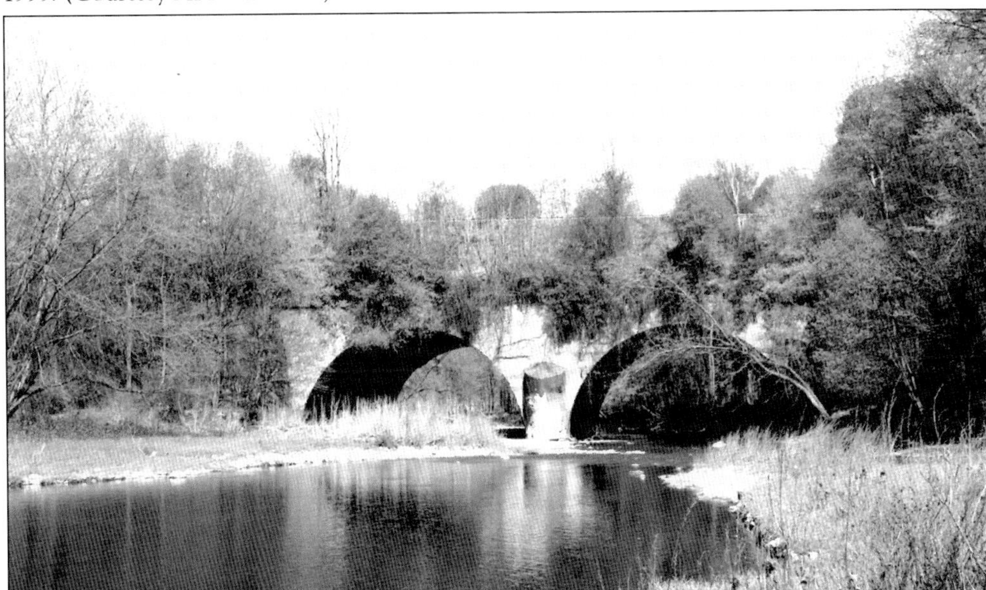

**Ashtabula Bridge Disaster Site, 2006.** The Ashtabula Bridge Disaster site in the wooded Ashtabula River gulf is accessed off an Indian Trails Park trailhead in Cederquist Park. Even 130 years removed, the site still yields artifacts from the tragic train wreck. A replacement wooden Howe Truss bridge was completed in January 1877. Pictured is CSX's c. 1904 causeway-styled four-track railroad bridge, constructed of concrete culverts and fill. (Author's collection.)

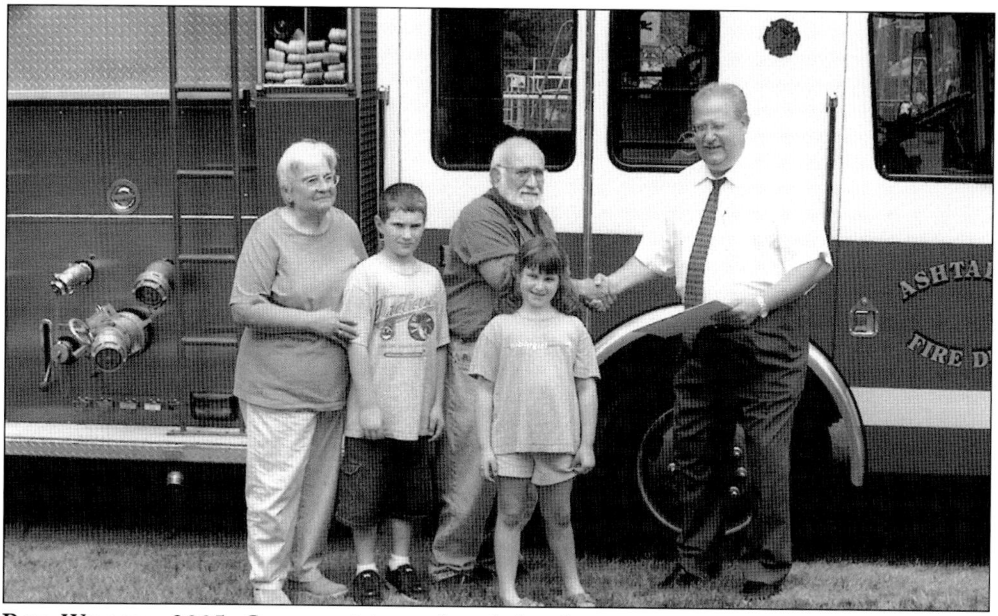

**BOB WILCOM, 2005.** Seen receiving a long-belated retirement proclamation from interim city manager Philip J. Varckette, retired 1959–1988 firefighter Capt. Bob Wilcom played an instrumental role in filling the department's historical archives. Wilcom donated old log books, firefighting equipment, photographs, and other memorabilia that he recovered during the 1975 demolition of the old headquarters Uptown Fire Station. Also pictured are Wilcom's wife Charlotte and grandchildren Erik and Lauren. (Courtesy AFD Archive.)

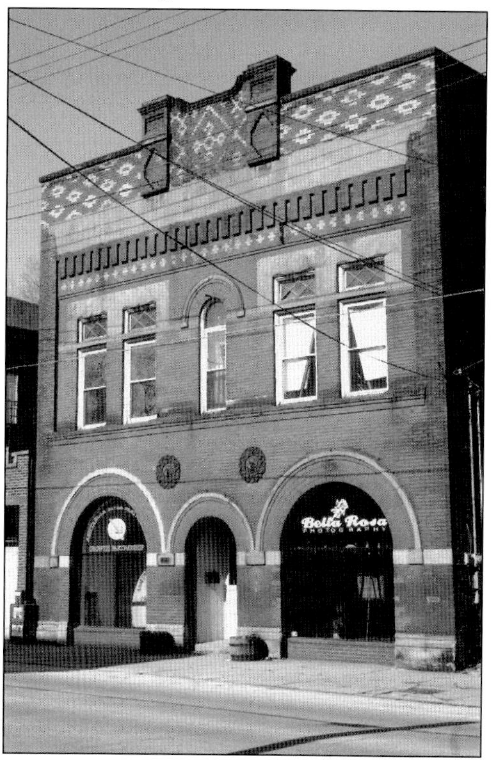


**HARBOR FIRE STATION, 1121–1123 BRIDGE STREET, 2006.** Prized for its ornate roofline, brick fancywork, soaring facade arches, and terra-cotta gargoyle medallions, Ashtabula's 1880s Harbor Fire Station today serves as an aesthetic anchor of Bridge Street's commercial revitalization. Expanded in 1906 with a 20-foot facade addition, the structure housed the volunteer Topky Fire Company and the Ashtabula Fire Department's paid professional Harbor Fire Station (1911–1959). (Author's collection.)

**UPTOWN STATION MEMORIAL.** When the Ashtabula Fire Department's c. 1912 Uptown Fire Station was closed in 1975 and razed for downtown urban renewal, several nostalgic furnishings were salvaged from the 63-year-old headquarters station and placed inside the newly opened Station One, 4326 Main Avenue. Salvaged items displayed in Station One's training room include Uptown's original brass Gamewell fire bell (left) and Seth Thomas pendulum clock (Courtesy AFD Archive.)

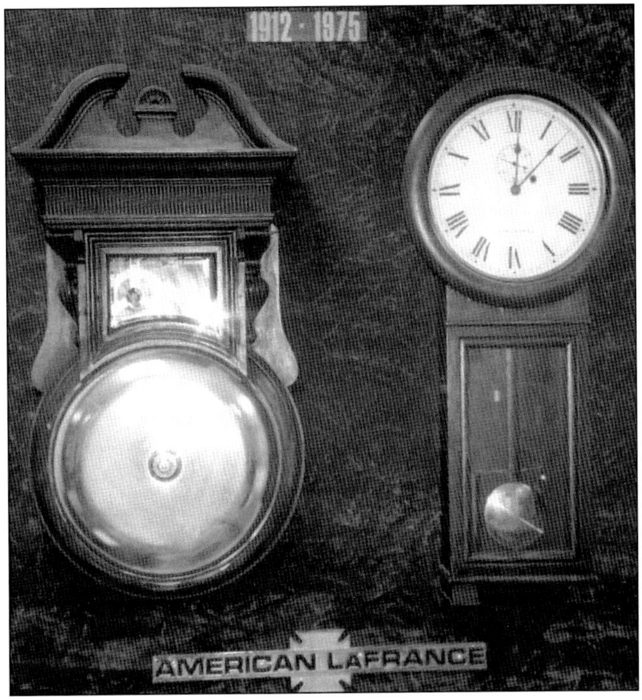


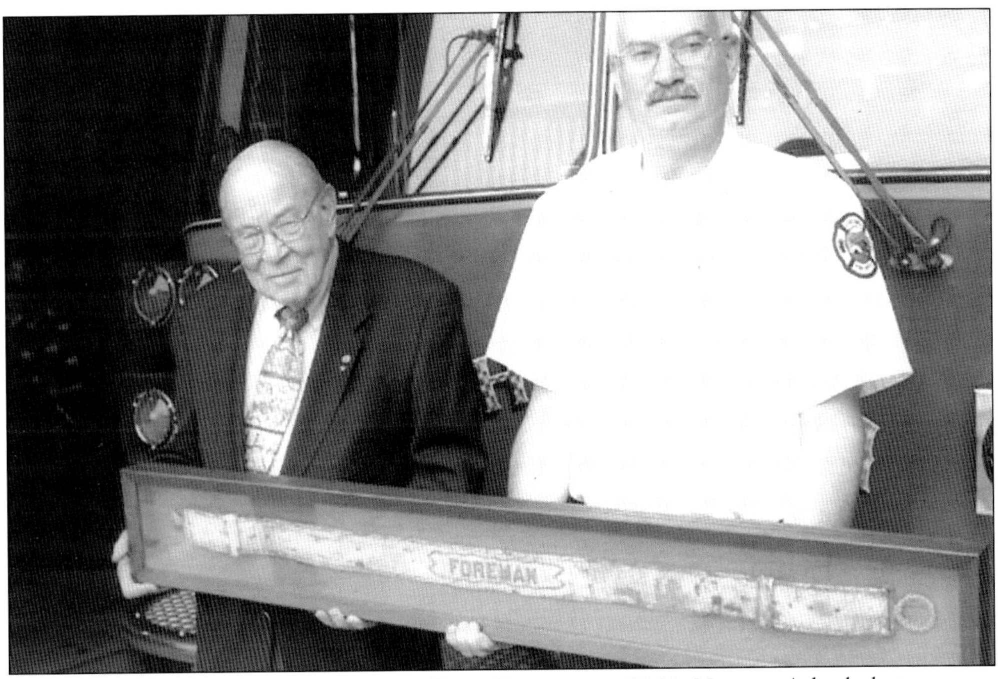


**PROTECTION FIRE COMPANY FOREMAN BELT DONATION, 2001.** Veteran Ashtabula insurance agent and local philanthropist Francis Herzog (left) presents Ashtabula fire chief Richard F. Balog Jr. with a Protection Fire Company "Foreman" uniform belt. The leather belt was crafted in Protection's red and white company colors. Long interested in local history, Herzog in 1975 funded the recovery of Lake Erie Hose Company's historic c. 1854 fire bell. (Courtesy AFD Archive.)

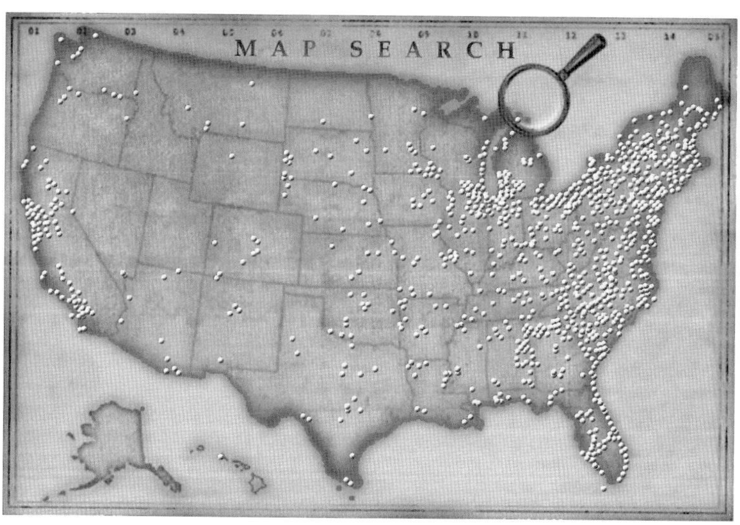
MAP SEARCH